Springer Theses

Recognizing Outstanding Ph.D. Research

For further volumes:
http://www.springer.com/series/8790

Aims and Scope

The series "Springer Theses" brings together a selection of the very best Ph.D. theses from around the world and across the physical sciences. Nominated and endorsed by two recognized specialists, each published volume has been selected for its scientific excellence and the high impact of its contents for the pertinent field of research. For greater accessibility to non-specialists, the published versions include an extended introduction, as well as a foreword by the student's supervisor explaining the special relevance of the work for the field. As a whole, the series will provide a valuable resource both for newcomers to the research fields described, and for other scientists seeking detailed background information on special questions. Finally, it provides an accredited documentation of the valuable contributions made by today's younger generation of scientists.

Theses are accepted into the series by invited nomination only and must fulfill all of the following criteria

- They must be written in good English.
- The topic should fall within the confines of Chemistry, Physics and related interdisciplinary fields such as Materials, Nanoscience, Chemical Engineering, Complex Systems and Biophysics.
- The work reported in the thesis must represent a significant scientific advance.
- If the thesis includes previously published material, permission to reproduce this must be gained from the respective copyright holder.
- They must have been examined and passed during the 12 months prior to nomination.
- Each thesis should include a foreword by the supervisor outlining the significance of its content.
- The theses should have a clearly defined structure including an introduction accessible to scientists not expert in that particular field.

B. Ambedkar

Ultrasonic Coal-Wash for De-Ashing and De-Sulfurization

Experimental Investigation and Mechanistic Modeling

Doctoral Thesis accepted by
Indian Institute of Technology Madras,
Chennai, India

Author
Dr. B. Ambedkar
Department of Chemical Engineering
Indian Institute of Technology - Madras
Chennai
India
e-mail: b_ambedkar@yahoo.com

Supervisor
Dr. R. Nagarajan
Department of Chemical Engineering
Indian Institute of Technology - Madras
Chennai
India
e-mail: nag@iitm.ac.in

ISSN 2190-5053
ISBN 978-3-642-25016-3
DOI 10.1007/978-3-642-25017-0
Springer Heidelberg Dordrecht London New York

e-ISSN 2190-5061
e-ISBN 978-3-642-25017-0

Library of Congress Control Number: 2011941669

Printed on acid-free paper

Springer is part of Springer Science+Business Media (www.springer.com)

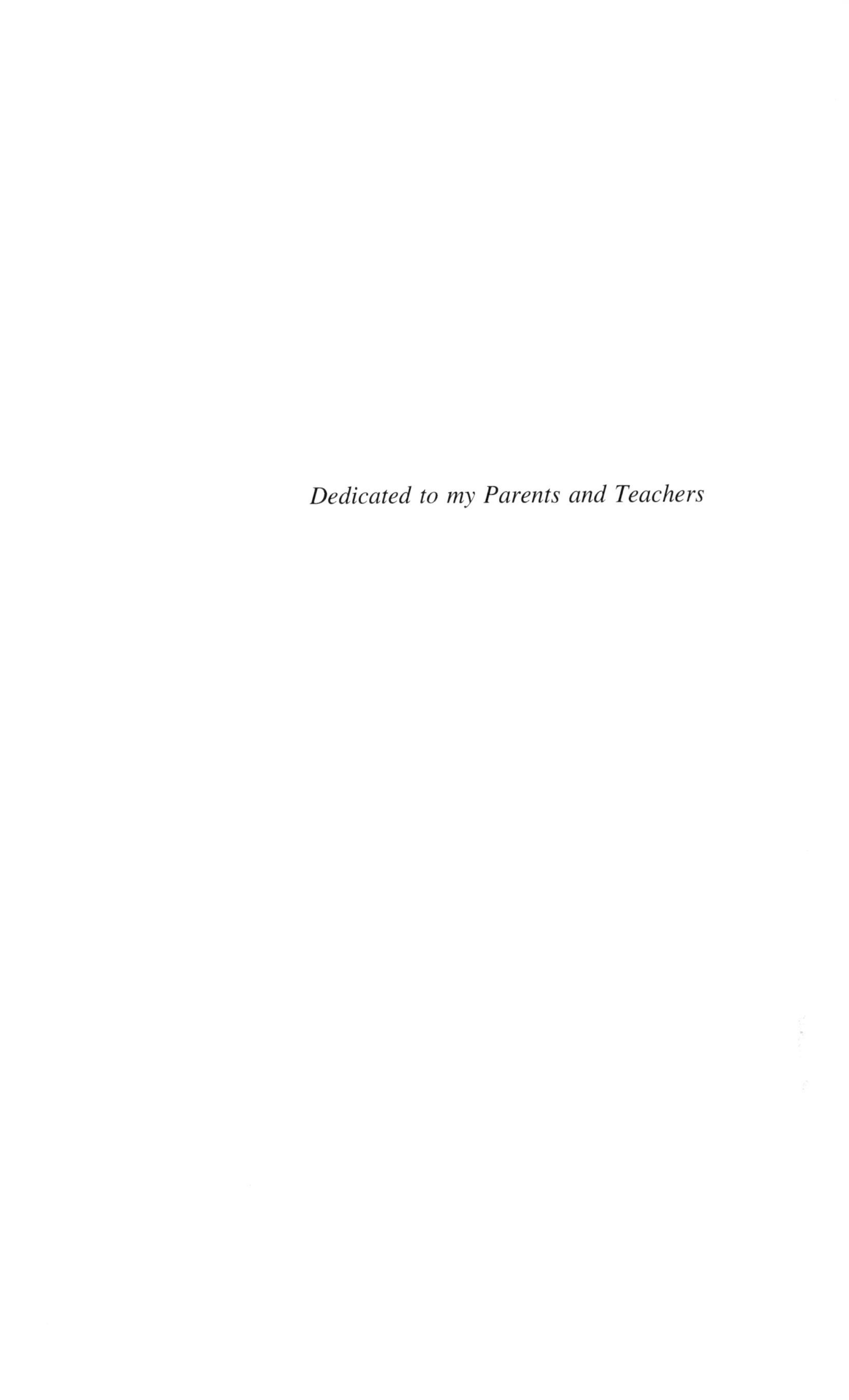

Dedicated to my Parents and Teachers

Parts of this thesis have been published in the following journal articles:

1. R. Nagarajan, B. Ambedkar, S. Gowrisankar and S. Somasundaram, "Development of Predictive Model for Fly-ash Erosion Phenomena in Coal-Burning Boilers", Wear, 267 (2009), 122–128.
2. B. Ambedkar, R. Nagarajan and S. Jayanti "Ultrasonic Coal-Wash for De-Sulfurization", "Ultrasonics Sonochemistry" 18, (2011), 718–726.
3. B. Ambedkar, T. N. Chintala, R. Nagarajan and S. Jayanti, "Feasibility of Using Ultrasound-Assisted Process for Sulfur and Ash Removal from Coal", "Chemical Engineering and Processing: Process Intensification" 50, (2011), 236–246.
4. B. Ambedkar, R. Nagarajan and S. Jayanti "Investigation of High Frequency, High-Intensity Ultrasonics for Size Reduction and Washing of Coal in Aqueous Medium", "Industrial Engineering and Chemistry Research" journal. (In Press)

Supervisor's Foreword

Dr. Ambedkar's thesis work has resulted in four papers, several conference presentations, and an "Outstanding Thesis" award. This comes as no surprise to me. Both in terms of the quantum of work and its quality, this dissertation is among the best in its breed. While "clean coal technology" is on everyone's wish-list, few have identified the relevant technologies. Removal of ash and sulfur from Indian coals prior to combustion using high-intensity acoustic fields is a novel concept with the promise of high throughput and cost efficiency. By means of well-designed experiments and sound analytical reasoning, Dr. Ambedkar has enunciated a path to follow. By focusing on Indian coals, this work becomes of immediate national relevance. While most Indian coals do have high ash content and low sulfur, there are exceptions; also, increasingly, foreign and Indian coals are being blended and used in power plants. Hence, simultaneous de-ashing and de-sulfurization become a critical need. Coal washeries being operated in India are unable to clean coal aggressively, being designed primarily to remove loosely-held dust, dirt and other debris. But more of these are being brought on-line regularly. This investigation paves the way for optimizing the design of coal washeries for maximum effectiveness with minimum cost and pollutant discharge. This is seminal work with potential for significant impact to the practice of coal washing in India.

Chennai, September 2011 R. Nagarajan

Acknowledgments

It is an honor for me to thank Prof. R. Nagarajan and Prof. Sreenivas Jayanti for their valuable guidance and constant support throughout my Ph.D. work. I profusely thank Prof. R. Nagarajan without whose support and continuous inquisition on the research topics, I wouldn't have completed this work. I thoroughly enjoyed the discussions that I had with him on various occasions. Apart from this, I inspired a lot from him in terms of professionalism and time management. I admired the prompt response from him in the form of suggestion, innovative ideas and the solution, the moment problem comes. I felt that, less than five minutes of discussion with him cracks the problem even it looks like a mountain. I thank Prof. Sreenivas Jayanti for his valuable guidance and discussions that I had with him have been very useful. I feel that I really need to learn a lot from him in terms of inquisition on research topics.

I also thank my doctoral committee members for their valuable suggestions. My special thanks to Prof. A. Kannan and Prof. S. Ramanathan for his valuable suggestions and encouragement.

I thank Prof. Krishnaiah, Dean of Academic Research, Prof. Pushpavanam, present HOD and Prof. Shankar Narasimhan, former HOD, for their constant support and encouragement.

I thank all my lab mates Dr. Jagannathan, Ms. Dhanalakshmi, Ms. Mercy, and Mr. Balakrishnan sharing things generously and encouragement. A special thanks to former lab mates Dr. Vetrimurugan and Dr. Gopi for his motivation.

My special thanks are due to Mr. Thirunavukarasu, Mr. Joseph and Mr. Palanivel for their help in using the workshop facilities and also thank Mr. Senthil Kumar (Project Attendant) for his help while doing experiments.

I sincerely thank my best friends Dr. A. Subramani (GE) and S. Anbudayanithi (CIPET) who accompanied me throughout my stay in the IIT campus and whose support, love, affection and constant motivation. I thank my friend Mr. Murugaraja (GE) and Mr. Thirunavukarasu (MRL) and Mr. JaiKumar (L&T) for their supporting and encouragement. I thank Dr. Ravi (DOE, US) for his prompt help in getting me the important journal articles.

A special thanks to Dr. Sivaji, Dr. Shyam, Mr. P.V. Suresh and Dr. Selvanathan (former DCF head) for his constant support, help and motivation. I thank all the research scholars of the chemical engineering department, Sandilya, Dr. Selvaraju, Sreenivasa rao, Dr. Sivakumar, Kanagasabai, Dr. Murugan, Vijay, Magesh Arigonda, Dr. Prasanna, Ashraf, Dr. Manivannan, K. Suresh, Noyal, Hemanth, Ram Satish, Prabhu, Venkatesan, and from Metallurgical department Kesavan, Gerald and many others for being very friendly during the period of my stay in the campus.

I thank all my esteemed school teachers and classmates for their sincere wishes; I would like to say my heartfelt thanks to Rev. J. P. James Paul and church believers for their continuous prayer and blessing. I also thank my best childhood friends Mr. Sivakumar and Mr. Krishnamoorthi for their support and good wishes.

Finally, I should thank my beloved parents Ms. Amsha and Mr. Balraj, for their constant support and encouragement throughout my studies so far. Without whose support I wouldn't have come this far as what I am today. I thank my brothers Mr. Alexander, Mr. Abraham and Mr. Kamesh for their support, unconditional love and affection on me. Last not least, my special thanks to my wife Ms. Puviarasi for her support in the form of love, affection and encouragement.

Contents

Nomenclature

Symbols

f	Frequency of ultrasound (kHz)
μ	Micrometer
Cs	Reagent concentration (mol/liter or gm/liter)
t	Sonication time (min)
A_{sp}	Total specific surface area (m^2/kg)
R_X	Rate of reaction with mass transfer (liter/mol*s)
d_{pi}	Initial coal particle size (μ)
d_{pf}	Final coal particle size (μ)
e_d	Energy dissipated/unit mass of liquid (W/kg)
To	Initial reaction mixture temperature (°C)
T_f	Final reaction mixture temperature (°C)
V_l	Reagent volume (m^3)
M_C	Mass of coal (Kg)
K_m	Mass transfer co-efficient (m/s)
D_{eff}	Effective diffusivity (m^2/s)
X_A	Fractional conversion of solid reactant
K	Model constant
a, b, c	Model parameters
E	Erosion loss (g)
d_p	Fly-ash particle size (μ)
Q_p	Fly-ash quantity (g)
V_F	Volumetric flow rate (m^3/min) [Area of Nozzle * Velocity]
Ro	Initial radius of particle (μ)
M_A	Molecular weight of sulfur

Abbreviations

AOP	Advanced Oxidation Process
ASTM	American Standard Testing Materials
DOE	Design of Experiments
EDAX	Energy Dispersive X-ray Analysis
FGD	Flue Gas De-Sulfurization
HCl	Hydrochloric Acid
HLA	High Level Analysis
HNO_3	Nitric Acid
H_2O_2	Hydrogen Peroxide
IS	Indian Standard
LPC	Laser Particle Counter
OH	Hydroxyl Radicals
ROM	Run of Mill
SEM	Scanning Electron Microscope
TSR	Total Sulfur Removal (%)
USCW	Ultrasonic Coal-Wash

Chapter 1
Introduction

1.1 Importance of Coal Washing

Coal is the largest source of fuel for generation of electricity throughout the world. The power sector will be the main driver of India's coal consumption. Currently, around 69% of the electricity consumed in India is generated from coal. Coal reserves in India are plentiful but of low quality. India has 10% of the world's coal reserves, at over 92 billion tons, third only to USA and China in total reserves [1]. The full utilization of coal is limited by the presence of high levels of ash and sulfur in it. During coal combustion, the mineral matter transforms into ash, and the amount of ash is so large that it is not easy to utilize ash effectively. Fly-ash is the finely divided mineral residue resulting from the combustion of ground or powdered coal in electricity generating plant. It consists of inorganic matter present in the coal that has been fused during coal combustion.

Mineral matter in coal causes several disadvantages, including: unnecessary cost for transportation, handling difficulties during coal processing, leaching of toxic elements during ash disposal, sulfur emission from pyrite-like minerals giving rise to an environmental problem, and ash deposition leading to the deterioration of boilers and accessories (thereby diminishing operating efficiency). During coal combustion, fly-ash particles entrained in the flue gas from boiler furnaces in coal-fired power plants can cause serious erosive wear on steel surfaces along the flow path, thereby reducing the operational life of the mild-steel heat-transfer plates that are used in the rotary regenerative heat exchangers. Moreover, in technical practice, erosion is often accompanied by a chemical attack.

In coal-fired power stations, nearly 20% of the ash in the coal is deposited on boiler walls, economisers, air-heaters and super-heater tubes and is eventually taken out as bottom ash. The deposited ash is subsequently discharged as slag and clinker during the soot-blowing process. The rest of the ash is entrained in the stream of flue gas leaving the boiler. The ash-laden flue gas passes through the narrow passages between the corrugated steel plates that constitute the air heater elements. The ash

B. Ambedkar, *Ultrasonic Coal-Wash for De-Ashing and De-Sulfurization*, Springer Theses, DOI: 10.1007/978-3-642-25017-0_1,

particles collide with the surfaces of the steel air heater elements and material is eroded from the surfaces. In advanced stage of erosion, the plates become perforated. The air heater elements fail once they cannot maintain their structural integrity.

SOx as a pollutant are a real threat to both the ecosystem and to human health. Sulfur is found in two forms in coal: (1) Inorganic sulfur, and (2) Organic sulfur. The inorganic sulfur again classified into two class (a) sulphate sulfur and (b) pyritic sulfur. Sulphate sulfur occurs in combination with Ca, Mg, Ba, Fe, etc. Pyritic sulfur in coal are pyrite (FeS_2), dimorphic marcasite (FeS_2), sphalerite (ZnS), galena (PbS), chalcopyrite ($CuFeS_2$), pyrhhotite ($Fe_{1-x}S$), arsenopyrite (FeAsS) and others. The chemical structures of organic sulfur components of coal are generally part of the macro-molecular structure of the coal itself. Organic sulfur is chemically bonded and very difficult to remove by physical cleaning methods. The organic sulfur components can be broadly divided into aliphatic and aromatic or heterocyclic sulfur structures.
Methods to control SO_2 emissions may be classified as:

1. De-sulfurization of coal prior to combustion (Physical, Chemical, Microbial),
2. The removal of sulfur oxides during combustion,
3. The removal of sulfur oxides after combustion, and
4. Conversion of coal to a clean fuel by gasification and liquefaction.

However, these are ineffective in the sense that time and energy consumption are high, and many chemicals are involved, introducing difficulties in handling of by-products during process. Nowadays, online flue gas de-sulfurization is being attempted to remove sulfur from coal post-combustion. The biggest disadvantage associated with this method is formation of by-products [Flue Gas Desulfurization (FGD) gypsum is one]. According to the American Coal Ash Association's annual Coal Combustion Product Production and Use Survey, total production of FGD gypsum in 2006 was approximately 12 million tons. Close to 9 million tons of FGD gypsum were put to beneficial use, while the remainder was land-filled. There is, at present, no way for effective usage of all FGD gypsum generated as by-product [2].

There are also concerns about environmental effects when FGD gypsum is used for soil amendment, and there are some reports on how chemical properties of soils, plants and animals are affected following FGD gypsum application (Environmental Protection Agency, USA). Concentrations of elements in soil, soil water, plant tissue and earthworms were measured. Results indicate that concentrations of Ca and S increased in plant tissue, soil, and soil water and the concentrations of Al and Fe decreased in plant tissue by gypsums. This will lead to acute and chronic effects to humans as well as plants. Over the next 10 years, annual production of FGD gypsum may double as more coal-fired power plants come online, and as scrubbers are added to existing power plants to meet the environmental clean-air standards. In the worst case, where sulfur in coal is 10% or higher, releasable sulfur amount can become very high. This would lead to unnecessary transport and storage before, as well as after, combustion in terms of FGD gypsum. There is clearly a need for removing ash and sulfur from coal prior to combustion. Ultrasonic coal-wash is one

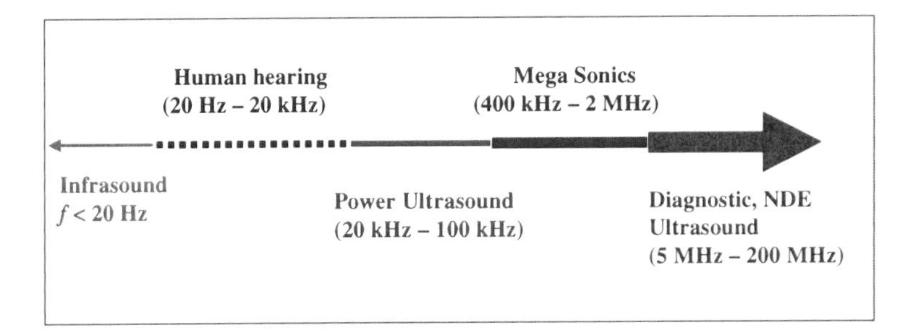

Fig. 1.1 Sound frequency range

such technique to effectively remove ash and sulfur from coal prior to combustion. In addition, it is easy to scale-up as a continuous process.

1.2 Ultrasound-Assisted Process Intensification

A wave is defined as a disturbance that propagates through space and time, usually with transference of energy. Waves travel and transfer energy from one point to another, often with no permanent displacement of the particles of the medium (that is, with little or no associated mass transport); they consist instead of oscillations or vibrations around almost fixed locations. A sound wave is a mechanical pressure wave that propagates or travels through a medium due to the restoring forces it produces upon deformation. Since a sound wave consists of a repeating pattern of high-pressure and low-pressure regions moving through a medium, it is sometimes referred to as a pressure wave.

Ultrasound is cyclic sound pressure with a frequency greater than the upper limit of human hearing. Although this limit varies from person to person, it is approximately 20 kHz (20,000 Hz) in healthy, young adults and thus, 20 kHz serves as a useful lower limit in describing ultrasound. The frequency ranges of sound are shown in Fig. 1.1. Sound wave of frequency less than 20 Hz is known as infrasound.

Ultrasound is a novel technology which is in widespread use in various scientific and medical fields [3, 4]—e.g., surface cleaning in microelectronics manufacturing, biomedical device cleaning, sono-chemical reactors designed to accelerate chemical reactions by several orders of magnitude, sono-intensification of mass-transfer and heat-transfer rates, sono-mixing and de-stratification in tall containers, nano-particle fabrication by sono-fragmentation, etc.

1.2.1 Basic Mechanism of Ultrasound

When ultrasound is applied to a medium such as water, the basic physical phenomena involved in producing changes observed (physical and chemical

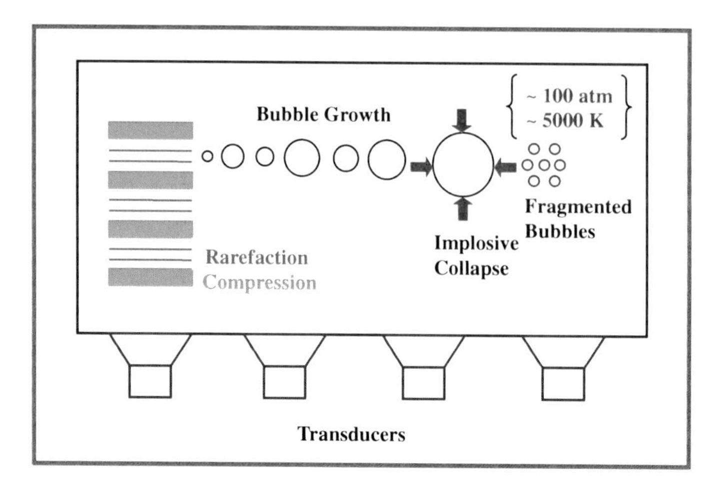

Fig. 1.2 Acoustic cavitation in ultrasonic tank

effects) in the medium are of two types—acoustic cavitation and streaming. Cavitation is the dominant mechanism in ultrasonic fields in the <100 kHz frequency. Two types of acoustic cavitations are identified, namely, stable cavitation and transient cavitation. In stable cavitation, bubbles continue to oscillate near their resonance size without collapsing; in transient cavitation, encountered in our system, bubbles grow and collapse, as observed visually and via measurements of cavitation intensity. Cavitational collapse results in extreme conditions producing light emission, shock waves, and localized high temperatures (up to approx. 5000 K) and pressures (up to 100 atm). These shock waves are responsible for the rupturing of neighboring solids (which may be vessel walls or immersed solids), leading to the generation of shear forces and eddies which, in turn, lead to an increase in turbulent energy dissipation. The number of these shock waves is related to the frequency of the waves [5, 6]. Typical acoustic cavitation that occurs in low-frequency ultrasonic tank is shown in Fig. 1.2.

Acoustic streaming refers to unidirectional flow currents in a fluid formed due to the presence of sound waves. Typical acoustic streaming that occurs in high-frequency ultrasonic and megasonic tanks is shown in Fig. 1.3. The formation of acoustic fountains is observed at the center of the transducer locations. Acoustic streaming comprises several important effects: (1) bulk motion of the liquid (Rayleigh streaming), (2) micro-streaming (Eckert streaming) and (3) streaming inside the boundary layer (Schlichting streaming). The primary effect of acoustic streaming is steady bulk motion of the liquid which generates shear force. A second effect of acoustic streaming is micro-streaming. Micro-streaming occurs at the substrate surface, outside the boundary layer, due to the action of bubbles as acoustic lenses that focus sound power in the immediate vicinity of the bubble. Micro-streaming aids in dislodging particles and contributes to megasonic

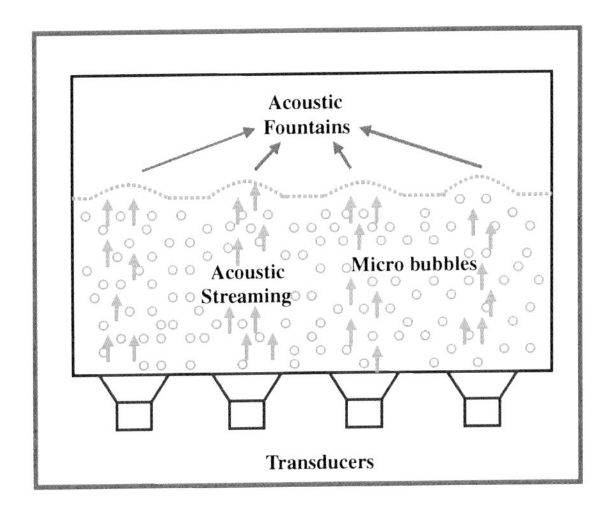

Fig. 1.3 Acoustic streaming in ultrasonic tank

cleaning. Most of the flow induced by acoustic streaming occurs in the bulk liquid outside the boundary layer. However, there is a third effect of acoustic streaming—Schlichting streaming—which is associated with cavitation collapse. Schlichting streaming occurs inside the boundary layer and is characterized by very high local velocity and vortex (rotational) motion. Acoustic streaming, both inside and outside the boundary layer, enhances cleaning and other chemical reactions.

Two basic mechanisms for acoustically enhanced coal washing have been suggested by Mason et al. [7]: (a) an increase in the abrasion of suspended coal in slurries leading to the removal of dust material from the surface of coal, and (b) an enhanced leaching of contaminants (mineral matter) from the interior of coal particles. Under the influence of ultrasound, normal leaching occurs, but several additional factors contribute towards improvements in the efficiency. These include:

1. Asymmetric cavitation bubble collapse in the vicinity of the solid surface, leading to the formation of high-speed micro jets targeted at the solid surface. The micro jets can enhance transport rates and also increase surface area through surface pitting.
2. Particle fragmentation through collisions will increase surface area.
3. Cavitation collapse will generate shock waves which can cause particle cracking through which the leaching agent can enter the interior of particle by capillary action.
4. Acoustic streaming leads to the disturbance of the diffusion layer on the surface.
5. Diffusion through pores to the reaction zone will be enhanced by the ultrasonic capillary effect.

The objective of the present study is to investigate the effectiveness of ultrasound-assisted washing for Indian coals which are distinctly different from others from the point of view of having a large amount of ash and very high sulfur.

1.3 Outline of Thesis

The remaining part of thesis is arranged as follows: Chap. 2 includes a detailed review of literature on effect of high-ash and high-sulfur coal burning within the boilers and environment, existing conventional reagent-based de-ashing and de-sulfurization methods and their demerits relative to ultrasonic methods, mechanism of ultrasound in aqueous medium and its contribution towards beneficiation of high-ash and sulfur coal. This chapter also includes scope for the present work and the specific objectives of present work. Assessment of the fly-ash erosion potential of Indian coals is discussed in Chap. 3. In Chap. 4, experimental studies on ultrasonic coal beneficiation are presented. This chapter includes studies on aqueous and reagent-based coal beneficiation. Experimental studies and optimization of reagent-based ultrasonic coal de-sulfurization is discussed in Chap. 5, where the effects of process parameters on TSR as well as development of mechanism-based model to predict TSR and scale-up of ultrasonic reagent-based coal de-sulfurization are presented. In Chap. 6, assessment of benefits from ultrasonic coal-wash is presented. It includes a proposed flow diagram for ultrasonic coal-wash on industrial scale, with projections on reduction of fly-ash impact metal erosion and corrosion-accelerated erosion of coal burning boilers due to minimization of particulate and SOx emission by ultrasonic coal-wash (USCW). Finally, summary, conclusions and recommendations are presented in Chap. 7.

References

1. www.worldcoal.org
2. www.epa.gov/osw
3. Nagarajan R (2006) Use of ultrasonic cavitation in surface cleaning: a mathematical model to relate cleaning efficiency and surface erosion rate. J Inst Environ Sci 49:40–50
4. Suslick KS, Hammerton DA, Cline RE (1986) The sonochemical hotspot. J Am Chem Soc 108:5641–5642
5. Suslick KS (1988) Ultrasound: its chemical, physical and biological effects. VCH, New York
6. Lorimer JP, Mason TJ, Fiddy K (1991) Enhancement of chemical reactivity by power ultrasound: an alternative interpretation of the hot spot. Ultrasonics 29:338–343
7. Mason TJ, Collings A, Sumel A (2004) Sonic and ultrasonic removal of chemical contaminants from soil in the laboratory and on a large scale. Ultrason Sonochem 11:205–210

Chapter 2
Literature Review

2.1 General

This review addresses work on effect of burning high-ash and high-sulfur coal within the boiler equipment and environment. Within the boiler equipment, erosion of boiler accessories due to coal fly-ash impaction is reviewed, along with: ultrasound-assisted particle breakage; application of ultrasound in various fields; some of the patented ultrasonic coal-wash process for de-ashing; existing conventional chemical-based de-sulfurization methods and their demerits relative to ultrasonic methods; mechanism of ultrasound in aqueous medium and its contribution towards high-sulfur coal de-sulfurization; and, ultrasound-assisted high-sulfur coal and high-sulfur diesel fuel de-sulfurization already initiated by other researchers.

2.2 Effect of Burning High-Ash and High-Sulfur Coal

2.2.1 Boiler Equipment

The main disadvantages of burning high-ash and high-sulfur coal within the boiler are erosion and corrosion-accelerated erosion of boiler accessories. Erosion is defined as a process by which material is removed from the layers of a surface impacted by a stream of abrasive particles. Erosion can be broadly classified as solid particle erosion, slurry erosion and cavitation erosion. When the particles strike the substrate, part of their kinetic energy is spent on creating new particles, part on indentation of substrate, and a part on rebounding. In case of brittle materials, erosive wear is predominant in case of normal impact, whereas in case of metals, maximum erosive wear occurs at shallow angles. If the striking particle is much harder than the substrate and the effect of the force on particle is large, abrasion predominates.

B. Ambedkar, *Ultrasonic Coal-Wash for De-Ashing and De-Sulfurization*,
Springer Theses, DOI: 10.1007/978-3-642-25017-0_2,

Finnie [1] proposed the first analytical erosion-model. This model included a variety of parameters that influence the amount of material eroded from a target surface and the mechanism of erosion. It was observed that the wear of a surface due to solid particle erosion depends on the motion of the particles in the fluid, as well as the behavior of the surface when struck by the particles. These two parts of the problem are related in that a surface, roughened by erosion, may increase the fluid turbulence, and hence, accelerate the rate of material removal.

Hutchings and Winter [2] studied the mechanism of metal removal by impacting the metal targets at an oblique angle by metal balls at velocities up to 250 m/s. They suggested that the initial stage of metal removal is the formation of lip at the exit end of the crater, caused by shearing of the surface layers. Above a critical velocity, this lip is detached from the surface by the propagation of ruptures at the base of the lip.

Jennings et al. [3] derived mathematical models based on target melting and kinetic energy transfer for predicting ductile target erosion. Dimensional analysis was employed in the development of a mathematical model for predicting the erosion of ductile materials. The model identified an erosion mechanism (target melting) which was verified in an erosion testing program using three stainless steels, two aluminium alloys, a beryllium copper alloy and a titanium alloy; the erosive agents were three dusts with hard angular particles, and one dust with spherical particles.

By extending the relations of Hertz and Raleigh, Soo [4] studied ductile and brittle modes of erosion by dust and by granular materials suspended in a gas moving at moderate speeds with conditions including directional impact, random impact, and sliding-bed motion. Their experimental results show that the ductile mode, which is typical of metal targets, is characterized by maximum erosion occurring at some intermediate incidence angle between 0 and 90°.

Foley and Levy [5] investigated the erosion of heat-treated steels. The testing was conducted at room temperature using aluminium oxide particles with an average size of 140 μm in an air stream. An attempt was made to characterize the erosion behavior as it relates to the mechanical properties obtainable in these alloys by conventional heat treatments. It was found that the ductility of the steels had a significant effect on their erosion resistance which increased with increasing ductility, and that hardness, strength, fracture toughness and impact strength had little effect on erosion behavior.

Sundararajan and Shewmon [6] proposed a correlation between the erosion rate and the thermo-physical properties of the target, for the erosion of metals by particles at normal incidence. This model employs a criterion of critical plastic strain to determine when the material will be removed. It was concluded that their new erosion model (localized model), rather than the fatigue-type model, predicts very well the experimentally observed rates of erosion. The effect of hardness on erosion rate was also investigated. The volume erosion rate for pure metals is inversely related to the static hardness. Such behavior can be rationalized on the basis of the fact that the melting point of a pure metal is directly proportional to its static hardness value. In the case of high-temperature erosion, there may be significant hardness effect on erosion, but in the case of room-temperature erosion, the influence is negligible.

Levy et al. [7] investigated elevated-temperature erosion of steels. The elevated-temperature erosion behavior of several commercial ferritic and austenitic steels was determined over a range of temperatures from room temperature to 900° C. Austenitic steels were determined to have lower erosion rates than ferritic steels, and their hardness had no correlation with their erosion rate.

Meng and Ludema [8] analyzed the origin, content and applicability of most wear models and equations in literature. Their work focuses on the need for new methods of wear modeling and offers recommendations on how to model the wearing process; the authors have found over 300 equations for wear and friction.

Wang [9] investigated the erosion-corrosion behavior of two steels and several thermal spray coatings due to impaction by fly-ash from a bio-mass fired boiler through laboratory tests using a nozzle-type, elevated-temperature erosion tester. They found that this bio-mass fired boiler fly ash had relatively high erosive effect due to its composition containing high concentrations of chemically-active compounds of alkali, sulfur, phosphorous and chlorine.

Xie and Walsh [10] measured the erosion of carbon steel by fly-ash and unburned char particles in the convection section of an industrial boiler firing micronized coal. Ash and char particles suspended in the flue gas entrained by the jet were accelerated towards the surface of the specimen under varying temperatures (450–650°). Changes in the surface were measured using a surface profiler. They observed that erosion was slowest at the lowest metal temperature, regardless of the jet gas composition; and, under the nitrogen jet, erosion increased with increasing temperature. They have presented a model for simultaneous erosion and oxidation which is consistent with the temperature and oxygenation dependencies of the erosion rate.

Hubner and Leitel [11] carried out investigations on an erosion-corrosion apparatus to investigate time behavior of corrosion-resistant high-alloy iron-base materials containing hard phases, and optimized the materials for increased wear resistance under complex stress conditions.

Oka et al. [12] investigated the impact-angle dependence of erosion damage caused by solid particle impact. Erosion tests were conducted using a sand-blast type erosion test rig which included shallow impact angles. The dependence of erosion rates on impact angle was characterized by type of metallic (Al, Pb etc.), plastic and ceramic material. Impact velocity increased the erosion rate, but did not affect the dependence of erosion behavior on the impact angle for the metallic materials. Impact angle dependence was simulated by a basic equation involving a trigonometric function both of impact angle and of material hardness.

Hussainova et al. [13] investigated the surface damage and material removal process during particle–wall collision of solid particles with hard metal and cer-met targets. Targets were impacted with particles over a range of impact velocities (7–50 m/s) at impact angle of 67°. The experimentally-observed variations of the coefficient of velocity of restitution as a function of the test material properties, impact velocity and hardness ratio were adequately explained by a theoretical model presented by them.

With a FLUFIX computer code, Lyczkowski and Bouillard [14] analyzed the behavior of six representative erosion models (comprising both single-particle and fluidized bed models) selected from literature. Energy dissipation models are developed, and are shown to generalize to the so-called power dissipation model used to analyze slurry jet pump erosion. They have demonstrated, by explicitly introducing the force of the particle on the eroding material surface, that impaction and abrasive-erosive mechanisms are basically the same.

Using three different power-station ash types, Mbabazi et al. [15] investigated the effect of ash particle impact velocity and impact angle on the erosive wear of mild-steel surfaces through experiments. The experimental data were used to calibrate a fundamentally-derived model for the prediction of erosion rates. This model incorporates the properties and motion of the ash particles as well as target metal surface properties.

In their work, Marcus and Moumakwa [16] investigated the long-term solid particle erosion of a range of oxide and nitride-fired SiC-based ceramics and alumina with the aim of reducing wear damage in power plants. They carried out experiments using 125–180 μm silica sand at shallow and high impact angles, using an in-house built erosion testing machine simulating real industrial conditions.

Das et al. [17] investigated the effect of temperature on the basis of the observation that the erosion rate at acute impingement angle increases significantly with temperature, suggesting that steel tends to show behavior more typical of a ductile material at elevated temperatures. The yield stress and temperature functionality has been derived through a polynomial approximation for various grades of steel on the basis of the available tensile property data. Erosion behavior at elevated temperatures has been incorporated through the derived functionality of the tensile property (yield stress) with temperature, along with appropriate modification of yield strength.

Vicenzi et al. [18] investigated the effect of fly-ash erosion on three different thermal-sprayed coatings produced by high velocity oxygen fuel (HVOF) under high temperature conditions by means of an apparatus that simulated real conditions.

Wang and Yang [19] developed a finite element (FE) model of erosive wear of brittle and ductile materials. The FE model was used to simulate the effect of impact angle and velocity, and of particle penetration, on the targets. The predicted results were found to be in good agreement with the experimental and analytical erosion models.

Such erosion, together with the processes of blocking, fouling and corrosion, shortens the service life of the air heater elements. Once this happens, the power station unit has to be shut down in order to replace the damaged air heater elements. The resulting penalty is not only the cost of replacing the elements but also the cost of stoppage of power production. It is desirable, therefore, to be able to predict the rate of erosion of the air heater elements in order to plan systematically for their maintenance to avoid forced outages.

2.2.2 *Environmental and Health Hazards of Coal Burning Power Plants*

The two major environmental concerns today arising from the use of high-ash and high-sulfur coal are: increase in atmospheric carbon dioxide levels and acid rain. It is believed that combustion has partially contributed to the increase in atmospheric carbon dioxide levels. Increased atmospheric carbon dioxide levels may result in warmer climates due to the "greenhouse effect". The increase in atmospheric carbon dioxide prevents heat from escaping from the earth, thus warming the atmosphere. The combustion of coal also appears to contribute to acid rain, although precise measures of the scope and seriousness of acid rain are not clear or well understood. Out of the entire US electric industry, coal-fired power plants contribute 96% of sulfur dioxide emissions (SO_2), 93% of nitrogen oxide emissions (NO_x), 88% of carbon dioxide emissions (CO_2) and 99% of mercury emissions. [http://www.greenpeace.org/raw/content/seasia/en/press/reports/coal-plants-a-greenpeace-brie.pdf]

2.2.2.1 Global Warming Due to Coal Combustion

Carbon dioxide emissions are the essential outcome of coal combustion in power plants. Carbon dioxide has been identified as a heat trapping gas; it retains the infrared radiations returning from earth to sun, thus causing the global temperature to rise. These impacts include melting of polar ice, rise in sea-levels and the consequent flooding of coastal areas. In addition, it may increase erosion of coastal lands, subjecting coastal buildings and their residents to increased risks of violent storms. Coal emits 29% more carbon per unit of energy than oil, and 80% more than natural gas. CO_2 represents the major portion of greenhouse gases. Over the last 30 years, the concentration of greenhouse gases in the atmosphere has increased by 30%.

2.2.2.2 Acid Deposition Due to Coal Burning

Bituminous coal used in most power plants contains small amounts of sulfur and nitrogen. Combustion of coal in power plants converts them to sulfur and nitrogen oxides respectively. These oxides, upon reaction with water in the atmosphere, result in precipitation of acid, sometimes also called "acid rain". Acid rain is often prevalent downwind from coal burning power plants, indicating the connection between acid formation and airborne emissions caused by coal-fired power plants. Acidification of lakes and streams results in decline of aquatic animal populations. In addition, crop damage, forest degradation, impaired visibility and chemical weathering of monuments are the major results of acid deposition. Furthermore, presence of acidic substances in air entails human health risks such as asthma and bronchitis. In 1997, pollution controls from power plants to reduce acid rain cost approximately $100 per ton.

2.2.2.3 Particulate Matter and Ground-Level Ozone

"Fine particles" are a mixture of a variety of different compounds and pollutants that originate primarily from combustion sources such as coal-fired power plants. Fine particles are of gravest concern because they are so tiny that they can be inhaled deeply, thus evading the human lung's natural defences. Power plants also emit fine carbon soot particles directly from their smokestacks. In 1999, coal plants directly emitted nearly 300,000 tons of fine carbon soot particles. These suspended particulates are dangerous for human health and may cause respiratory illnesses. The airborne nitrogen oxide emissions associated with coal burning cause urban smog, which is a respiratory irritant. Moreover, increased ground level ozone due to nitrogen oxides reduces agricultural and commercial forest yields. 30,000 deaths each year are attributable to fine particle pollution from U.S. Power plants. It is further stated that hundreds of thousands of Americans suffer from asthma attacks, cardiac problems and upper and lower respiratory ailments associated with fine particles from power plants.

2.2.2.4 Mercury Emissions from Coal Burning

Most coal-fired power plants are major mercury emitters; mercury is present in coal in small traces and is released to the atmosphere during combustion. Although it is emitted in a non-hazardous elementary form, its accumulation in the environment can be hazardous for humans and wildlife. It is a neurotoxin, and if deposited in an aquatic environment in the form of methyl mercury, it can accumulate in invertebrates and fish and may affect their neural tissues. Coal-fired power plants are the single largest source of mercury pollution in the US. According to the US *National Wildlife Federation* (NWF), a single 100 MW coal-fired power plant emits approximately 25 pounds of mercury a year. According to the *US Centre for Clean Air Policy*, 50% of the mercury emitted from coal-fired power plants can travel up to 600 miles from the power plant. In 1994, mercury emissions by coal plants in the US reached 51 tons. According to NWF, as little as 0.002 pounds of mercury a year can contaminate a 25 acre lake to the point where fish are unsafe to eat.

These studies and data make clear that "clean coal technology" must be developed if coal continues to be used as a major source of energy. Removal of ash and sulfur from coal and coal combustion products is therefore a crucial component of clean coal technology.

2.3 Conventional Methods for De-Ashing and De-Sulfurization

Several methods are reported in literature for removal of mineral matter, total sulfur and different forms of sulfur from coal. The processes of sulfur removal from coal prior to combustion can be subdivided into physical and chemical methods. Of these, physical methods can remove the soluble sulphates, as well as a considerable portion of the coarse pyrite (separable by a magnetic separator), but the fine pyrite (tightly bonded with coal matrix) and organic sulfur remain largely untouched. On the other hand, many of the chemical methods can remove almost all of the pyritic sulfur and at least a portion of the organic sulfur. A number of chemical methods have been presented in literature. Table 2.1 lists the reagents whose use has been reported in literature and their effects on coal desulfurization. From Table 2.1, it is apparent that researchers are still looking for ways to maximize coal ash and sulfur removal, and to optimize the operating conditions by choosing suitable reagents.

Yuda and Ayse [40] investigated the effect of supercritical ethyl alcohol/NaOH on the solubilization and de-sulfurization of Beypazari lignite. Supercritical experiments have been done in a 15 ml micro reactor at 245° C for 60 min, by changing the ethyl alcohol/coal ratio from 3 to 20 under a nitrogen atmosphere. Increase in ethyl alcohol/coal ratio increased the yield of solubilization and de-sulfurization. Higher yields of extraction in the case of ethyl alcohol/NaOH experiments may be due to the fact that alcohols can transfer hydrogen more easily in the presence of bases. As the ethyl alcohol/coal ratio was increased from 3 to 20, the sulfur content of the coal decreased to 0.75%.

Li and Guo [41] have studied de-sulfurization of a high-rank coal using alcohol/KOH and alcohol/water under supercritical conditions in a semi-continuous reactor and in a batch reactor. In the semi-continuous reactor mode, it was found that supercritical de-sulfurization is mainly taking place within about one hour at 400° C. Ethanol/KOH solution as supercritical solvent enhanced the de-sulfurization process in which inorganic sulfur was removed. The reaction between ethanol and KOH takes place in three steps: Ethanol reacts with KOH and forms potassium ethoxide and water. Then, these two combine to form potassium ethanolate and hydrogen. Part of potassium ethoxide gives ethylene and KOH. With increasing KOH concentration, a large amount of hydrogen is produced and is absorbed by the coal. The effect of hydrogenation makes the radical fragment more stable.

Mukherjee et al. [34] have investigated de-mineralization and de-sulfurization of high-sulfur coals from Assam (India) using aqueous NaOH followed by HCl treatment. They found that compared to alkali and acid alone, successive treatments with alkali and acid resulted in significant removal of mineral matter and sulfur from the coal.

Charatuwai et al. [42] have studied de-sulfurization of Mae Moh (Thailand) coal with supercritical ethanol/KOH in a semi-continuous reactor. A two-level factorial design was applied, and process variables investigated were reaction temperature, pressure, and reaction time and KOH concentration. The effects of process variables on coal yield, as well as on ash reduction and total sulfur reduction, have been analyzed using analysis of variance. Among the four variables, temperature and KOH concentration were found to be significant factors for removal of total sulfur.

Mukherjee and Borthakur [43] have investigated the effect of leaching on high-sulfur sub-bituminous Assam (India) coal using KOH and acid on removal of mineral

Table 2.1 Various reagents used to remove ash and sulfur from coal

Author	Reagents used	Time	Sulfur and or ash removal
Steinberg et al. [20].	O_3 and O_2	1 h	Using a flow rate of 200 ml/min, 1% O_3 at 25 °C, 20% sulfur removed
Aarya et al. [21]	NaOH	8 h	Using 100 g/dm^3 NaOH at 80 °C, 30% sulfur removed 29% ash removed
Chandra et al. [22].	Atmospheric oxidation	106 days	44% sulfur removed (36% organic sulfur removal)
Krzymien [23]	Aqueous $CuCl_2$	48 h	Using 10 ml of 10% (vol) $CuCl_2$ at 200 °C, 100% sulfur removed
Chaung et al. [24]	Combination of dissolved oxygen and alkalis $NaHCO_3$, Na_2CO_3 and Li_2CO_3	1 h	0.2 M alkali solution with 3.4 atm O_2 partial pressure at 150 °C: Na_2CO_3: 72% of sulfur removed Li_2CO_3: 73.1% of sulfur removed At 0.4 M $NaHCO_3$: 77% of sulfur removed
Yang et al. [25]	NaOH	60 min	Using 10 wt% NaOH at 250 °C: 55% sulfur removed (95% pyritic and 33% organic sulfur removed)
Kara and Ceylan [26]	Molten NaOH at different temperatures	30 min	Using 20 wt% NaOH at 450 °C: 83.5% sulfur removed 91% ash removed from Dadagi lignite
Ahnonkitpanit and Prasassarakich [27]	Aqueous H_2O_2 and H_2SO_4	2 h	Using 15% H_2O_2 and 0.1 N H_2SO_4 at 40 °C: 48.7% total sulfur removed (97% pyritic, 89% sulphate and 7.1% organic sulfur removed) 72.2% ash removed
Ozdemmir et al. [28]	Chlorine in $CCl_4 + H_2O$	6 h	Using 0.033 l/min chlorine flow rate at ambient temperature and pressure: All pyritic and sulfate sulfur removed and 30% organic sulfur removed 15% ash reduced
Ali et al. [29]	H_2O_2, NH_4OH, $K_2Cr_2O_7$ and CH_3COOH	30 min	50–90% of sulfur removed, depending on concentration and solvent 50–55% of mineral matter removed, depending on concentration and solvent
Prasassarakich and Thaweesri [30]	Sodium benzoxide	90 min	Using 600 ml sodium benzoxide at 205 °C, 45.9% sulfur removed (83.7% sulphate, 68.6% pyritic, 33.3% organic sulfur removed)

(continued)

Table 2.1 (continued)

Author	Reagents used	Time	Sulfur and or ash removal
Rodriguez et al. [31]	HNO_3	2 h	Using 20% HNO_3 at 90 °C, 90% inorganic and 15% organic sulfur removed
Hamamci et al. [32]	Acidic Fe $(NO_3)_3 \cdot 9H_2O$	12 h	Using 50 ml of 1 M solvent at 70 °C, 72.2% sulfur removed (96.6% pyritic sulfur removed)
Aacharya et al. [33]	Thio-bacillas ferro-oxidants	30 days	91.81% sulfur removed from lignite 63.17% sulfur removed from polish bituminous coal 9.41% sulfur removed from Assam coal
Mukherjee and Borthakur [34]	H_2O_2 & H_2SO_4	4 h	Using 15% (vol) H_2O_2 and 0.1 N H_2SO_4: 45% of total sulfur removed (complete removal of inorganic sulfur and 31% removal of organic sulfur) 45% ash removed
Ratanakandilok et al. [35]	Methanol/water and methanol/KOH	90 min	Using 2% methanol and 0.025 g KOH/g coal at 150 °C: 58% total sulfur removed (77% sulfate, 47% pyritic and 42% organic sulfur removed) 24% ash removed
Sonmez and Giray [36]	Peroxy acetic acid	72 h	45% sulfur removed from Gediz lignite 85% sulfur removed from Cayirhan lignite
Aacharya et al. [37]	Aspergillus	10 days	78% sulfur removed with 2% pulp density
Baruah et al. [38]	Water	120 h	77.59% pyritic sulfur removed with aqueous leaching at 45 °C
Liu et al. [39]	Aeration + NaOH, HCl	5 h	Using 0.25 M NaOH at 90 °C with aeration rate of 0.136 m^3/hr and 0.1 N HCl solution 73% organic sulfur removed 83% sulfide sulfur removed 84% pyritic sulfur removed

matter and sulfur at temperatures of 90 and 150 °C. They reported that at 150 °C, successive treatments of coal with 18% KOH and 10% HCl leads to 52.7% desulfurization (all inorganic sulfur and 37% organic sulfur were removed from coal).

Mukherjee and Borthakur [44] have investigated the effect of mineral acids (HCl, HNO_3 and H_2SO_4) on de-mineralization of sub-bituminous high-sulfur Boragolai (Assam, India) coal at varying stirring speeds, and in the temperature range from ambient to 95 °C. They reported that HCl is less effective for de-mineralization compared to H_2SO_4 and HNO_3. They observed that de-sulfurization increases with increase in HCl concentration, and that increase of temperature to 95° C increases de-mineralization.

Alam et al. [45] have investigated the effect of process parameters on desulfurization of Mezino coal by HNO_3/HCl leaching. The parameters considered were reaction time, acid concentration, temperature and stirring speed. To optimize experimental parameters, Taguchi orthogonal experimental design was used with the chosen parameters. ANOVA indicated that acid concentration had the dominant effect on desulfurization.

The biggest disadvantage associated with conventional de-sulfurization is the processing time as well as increased reagent consumption. Also, at the end of the treatment, the coal is contaminated by byproducts that are produced during reaction. This may require neutralization of the treated coal. This indicates that there is a clear need for the new technology to overcome these issues.

2.4 Ultrasound-Assisted Coal Particle Breakage and Application of Ultrasound in Various Fields

This review comprises of four parts: particle breakage mechanism, application of ultrasound in various fields, some of the patented ultrasonic coal-wash process, and studies on free radical formation in an ultrasonic field.

2.4.1 Ultrasound-Assisted Coal Particle Breakage

Fridman [46] proposed a mechanism which explains the interaction of a cavitation bubble and a particle of the material medium, and also the effects of dispersion and coagulation in a cavitating liquid medium. It has been suggested that either cavitation or coagulation can be obtained by changing the physico-chemical and acoustic parameters of the medium. For more efficient dispersion, a combination of alternating cavitation and mechanical dispersion was recommended. The proposed mechanism was verified experimentally by slow-motion photography.

Kusters et al. [47] developed a model to describe the fragmentation of agglomerate powders by ultrasonication. An expression has been derived for the agglomerate fragmentation rate as a function of power input, suspension volume and agglomerate size. From the evolution of the mass mean diameter, it followed that the fragmentation rate varies linearly with agglomerate size, in agreement with experimental data. Fragmentation by erosion results in a bimodal fragment size distribution, requiring much finer section spacing in the sectional model than conventional fragmentation. In the case of an erosion-dominant type of fragmentation process, an additional term had to be included in the conventional breakage distribution expression to describe the production of fines. The amount of fines produced was found to be proportional to the surface area of the agglomerates. The fragmentation rate expression is evaluated by comparing simulated with experimental size distributions. The required time and energy for particle size reduction is calculated as a function of ultrasonic power input.

Kusters et al. [48] have investigated the energy requirement for sono-fragmentation. Ultrasonic field is extensively used to disperse submicron agglomerated powders in liquid suspensions. Experiments were conducted to illuminate the effect of suspension volume on the ultrasonic fragmentation rate. The fragmentation or grinding rate is inversely proportional to suspension volume. The reduction ratio increases with time faster at the small than at the large suspension volume for equal power input. Lower power input for ultrasonication favors efficient energy use. For eroding powders (e.g., silica, zirconia), the energy expenditure per unit powder mass (specific energy) by ultrasonic grinding is lower than that of conventional grinding techniques. In contrast, it is slightly higher than ball milling for non-eroding powders (e.g., titania).

Gopi and Nagarajan [49] investigated fabrication of alumina nano particles by sono-fragmentation. Breakage was more predominant in a low-frequency (<60 kHz) ultrasound field. It produces very fine particles, thereby increasing total surface area. The sphericity of the particle also increases with sonication time, due to the associated micro-polishing mechanism.

Raman and Abbas [50] have investigated the effects of intensity of ultrasound on particle breakage in liquid medium and also the effect of sonication power, temperature and contact time on particle breakage. The experiments were conducted at three different input power levels of 150, 250 and 350 W (amplitude ratios of 0.3, 0.5 and 0.7 respectively). The particle size in the form of mean chord length (l_m) was monitored in-line while the sonication was performed. A decrease in l_m was observed for all input power levels. Three different flow rates were studied: 1.0, 1.6 and 2.2 l/min. Breakage was more predominant at lower flow rates corresponding to larger values in residence time. As the residence time increases, the particles spend more time in contact with the breakage forces of the HIU (high intensity ultrasound) field.

Temperature has a significant effect on the cavitation phenomenon, which, in a liquid medium, is affected by its surface tension, viscosity and vapor pressure. Increasing temperature results in the liquids cavitating at lower intensities. This can be attributed to the increase in vapor pressure of the liquid, decrease in surface tension, and reduced viscosity of the liquid medium. The decrease in viscosity decreases the magnitude of the natural cohesive forces acting on the liquid, and thus, decreases the magnitude of the cavitation threshold. Lower cavitation thresholds translate into ease of cavity formation, thereby making higher temperatures more favorable for particle breakage. This is the reason for an increase in breakage of particles as temperature is varied from 10 to 25° C. As temperature is increased beyond 25° C, a decrease in particle breakage is observed. This is primarily caused by the cushioning effect of increased cavity internal vapor pressure at higher temperatures. Due to this cushioning effect, the intensity of the collapse, and subsequently the breakage, decreases above 25° C. There are, thus, two opposing factors that are at play during particle breakage; the first being the increase in the number of cavitation events with increase in temperature due to which particle breakage increases, and the second being the cushioning effect of the cavity internal vapor pressure, which has a suppression effect on the cavitation intensity and subsequently on particle breakage.

2.4.2 Application of Ultrasonic Process

Newman et al. [51] suggest a methodology for the remediation of soils contaminated with inorganic pollutants. Copper oxide-doped granular pieces of brick were used as a model for contaminated soil. By passing water across the substrate on an ultrasonically-shaken tray, a 40% reduction in copper content was achieved, whereas in conventional washing, only 6% reduction is realized. The majority of the copper was removed as a result of the removal of surface materials which were more heavily contaminated with the copper oxide.

Kruger et al. [52] investigated the effect of ultrasound on degradation of highly-volatile chlorinated compounds present in groundwater. The main constituent of high volatility, chlorinated hydrocarbon 1, 2-dichloroethane (1, 2-DCA), was taken as a model pollutant, and it was experimentally observed that the destruction rate of 1, 2-DCA in deionized water depends on intensity of ultrasound, initial concentration and sample volume. The highly-volatile chlorinated hydrocarbons are completely destroyed in natural ground water within 60 min, but all minor halogenated components are destroyed within 30 min of sonication time. It was also observed that the destruction rate of 1, 2-DCA in deionized water is independent of temperature, and the pH value of the 1, 2-DCA solution in deionized water decreases with sonication time.

Farmer et al. [53] reviewed the application of power ultrasound to surface cleaning of silica and heavy mineral sands. They conducted experiments which revealed that reducing the iron contamination due to surface coating of the silica grains from 0.025 to 0.012% Fe_2O_3 would make this sand suitable for the production of tableware glass, and also investigated the effect of ultrasonic power level and concentration of reagents (sodium carbonate) on iron reduction. The optimum concentration of reagents which reduces exposure time required to reach the maximum iron reduction was evaluated.

Kim and Wang [54] investigated the effectiveness of ultrasound in enhancement of soil flushing method. The degree of enhancement varies with many factors, such as soil type, soil density, flow rate, temperature, wave frequency, energy level and others. The test soils were Ottawa sand, a fine aggregate, and a natural soil; the surrogate contaminant was Crisco Vegetable Oil. The percent contaminant removal increases with increasing sonication power to a maximum around 100 W, then decreases. The contaminant removal at 140 W is about equal to that around 75–85 W, corresponding to a loss factor of about 1.8. The drop in contaminant removal beyond about 100 W can be attributed to the effect of cavitation. When cavitation occurs, the sound pressure level at a distance drops, because cavitation takes power away from the field. Therefore, cavitation can reduce the effective sonication power in the soil.

The influence of hydraulic condition was investigated under 100 W sonication power using three levels of hydraulic gradient: 1.6, 5.5, and 13.0. Percent contaminant removal was studied as a function of hydraulic gradient for both with and without sonication conditions. It is seen that the percent contaminant removal decreases with increasing hydraulic gradient. Increasing hydraulic gradient will increase discharge velocity and flow rate, other factors being equal. The contaminant in a soil with a higher void ratio can be removed more easily than a soil with a lower void ratio; also, the effect of soil density on contaminant removal seems to be less significant for the case with sonication than without sonication. The effectiveness of sonication in contaminant removal is greater at lower discharge velocity. This can be attributed to the relatively longer time for interaction between sound wave and contaminant under slower flushing.

Mason et al. [55] have investigated sonic and ultrasonic removal of chemical contaminants from polluted soil in the pilot-scale and on the large scale. They analyzed three different industrial-site polluted solids, namely DDT, PCB, and PAH- doped soil, and concluded that % removal of chemical contaminants was 75, 75 and 85% during 5, 30, and 60 min of ultrasonic washing using 20 kHz ultrasound and 200 g contaminated soil in 200 g water. Cooke et al. [56] reported that "ultrasound (0.455–1.46 W/cm^2) can extract at least 58% of mobile organic matter without rupturing any chemical bonds. The average molecular weight of the extract is 340–1,055".

2.4.3 *Patented Ultrasonic Coal-Wash Process for De-Ashing*

In a patented process and apparatus for treatment of flowing slurries of particulate material mixed in liquid (US Patent # 4741839 [57]), a wide, elongated, downwardly-slanted metal tray with upturned edge flanges, is cable-suspended for unrestrained vibratory flexing and undulation. The tray is provided with a plurality of ultrasonic transducers mounted on its underside, and the flowing slurry is delivered to the upper tray end, flowing lengthwise down the tray in a shallow flowing sheet. Ultrasonic vibratory energy coupled through the tray to the flowing slurry has a "microscopic scrubbing" action on all particles and agglomerates, breaking the surface tension on the particle, cleaning particle surfaces, and separating different constituent particles and coatings of gels, slimes, algae, clay or mud. Mixtures of fine particles of coal or other valuable minerals with ash, clay, rock or sand particles are separated with unexpected efficiency by these techniques. Advanced Sonic Processing Systems (Oxford, CT, USA; [www.advancedsonics.com]) offers a "Vibrating Tray Equipment Series", which is a high-volume ultrasonic trough effective in accelerating the surface dynamics of the fluidized particles. The ultrasonic cavitational energy scrubs each particle's surface as it flows over the tray. The cleaning effect produced by water alone is very effective in removing surface contaminates from the particulate pores. Chemical additives, added prior to the ultrasonic vibrating tray, become highly reactive in the acoustic field. This allows ore refining techniques to produce higher yields with lower consumable costs.

DOE, USA [www.rexresearch.com] reports on ultrasonic activation of several coal cleaning processes that in all cases "sonication demonstrated effects that would translate in production to processing efficiencies and/or capital equipment savings. Specifically, in the chlorinolysis process, pyritic S was removed 23 times faster with ultrasonic than without it. In NaOCl leaching, the total S extraction rate was three times faster with ultrasound. Two benefits were seen with oxy-desulfurization: ultrasonics doubled the reaction rate, and at slightly accelerated rates, allowed a pressure reduction from 960 to 500 psi".

Another ultrasonic process for cleaning coal has been patented in Great Britain (*British Patent # GB* 2,139,245 [58]); in this, coal slurry (pH 6–9) is agitated with ultrasound and separated by centrifuging or froth flotation. A second treatment with ultrasound and ozone releases more contaminants. There is another US patent (*US Patent* # 4,156,593 [59]) on an ultrasonic coal cleaning process in which coal contaminants (e.g., pyrites, clay) are removed from coal slurry at relatively low temperature and pressure and at increased throughput rates by an ultrasonic source. Pyrites are reduced from ~ 30 to $\sim 0.7\%$.

There are many such references in literature to the employment of ultrasonics in coal cleaning. However, none of them contains a systematic study of the effect of ultrasonic field parameters (amplitude, frequency), nor a delineation of the mechanisms involved. The study reported in this thesis focuses on these aspects.

2.4.4 Ultrasound in Aqueous Medium

Ultrasound is cyclic sound pressure with a frequency greater than the upper limit of human hearing. It starts from the frequency of 20 kHz. Ultrasound behaves differently in liquid and liquid–solid media compared to gas medium. Ultrasound in aqueous medium produces highly reactive species such as OH radicals, H_2O_2 and ozone that are strong oxidizing agents of high oxidation potential (2.8, 1.8 and 2.1 V respectively). These radicals are capable of initiating and enhancing oxidation and reduction reactions. Oxidation occurring due to ultrasound is called "advanced oxidation process" (AOP). Sonication enhances mass transfer and chemical reaction, and is expected to reduce or eliminate chemical usage, resulting in minimum disposal problems. Lindstorm and Lamm [60] first suggested the mechanism for this reaction, followed by many researchers who proved it in different manners by experiments. Webster [61] explained the cavitation mechanism as follows: Two classes of chemical effect are induced by ultrasonic cavitation. The first is the acceleration of reactions, and the second class of effect is the initiation of reactions that would not otherwise occur; this takes place predominantly in an aqueous medium. Under the action of cavitation, water decomposes into free radicals.

$$H_2O \rightarrow H^+ + OH^-$$

The predominant back reactions attendant on this process are

$$OH + OH \rightarrow H_2O_2; \quad H + H \rightarrow H_2$$

The products of these reactions are then responsible for secondary reactions involving dissolved substances. The reacting ions or molecules will be selectively subjected to reduction or oxidation according to their properties and structure. The oxidation of dissolved substances is detectable in the absence of dissolved oxygen. In its presence, the rate of formation of hydrogen peroxide is increased, with a consequent increase in the rate of oxidation; this effect has been attributed to the occurrence of the reaction

$$H + O_2 \rightarrow HO_2 \quad \text{followed by} \quad HO_2 + HO_2 \rightarrow H_2O_2 + O_2$$

Makino et al. [62] reported that intense ultrasound causes chemical damage through the phenomenon called cavitation. Cavitation produces high local instantaneous temperatures, pressures, and sonoluminescense. In sonolysis studies of aqueous solutions, it is proposed that hydroxyl radicals (OH^-) and hydrogen atoms (H) are produced by ultrasound. Riesz et al. [63] were able to observe by spin traps the highly-reactive radicals produced during cavitation. Christman et al. [64] found experimental evidence for free radicals produced in aqueous solutions by using electron spin resonance method (ESR). Misik and Riesz [65, 66] conducted spin trap and electron spin resonance studies to investigate free-radical formation and sonochemical reactions in organic liquids using 50 kHz frequency of ultrasound. Margulis [67] proposed that the fundamental problem in sonochemistry and cavitation is that hot-spot theory is not sufficient to elucidate the mechanism involved. A new electrical theory has been proposed and validated with experiments, with the electric field developed during cavitation mechanism being identified as a contributor for enhancing the sonochemical reaction. Entezari and Krus [68] conducted an experiment to explain the effect of frequency on sono-chemical reactions. The effect of sonication on iodide oxidation in presence of an air and argon atmosphere using two extreme frequencies (20 and 900 kHz) was investigated. The rate of sonochemical oxidation in an aqueous solution is about three times faster in an air atmosphere compared to argon environment. The H^+, OH^- and H_2O_2 produced by the ultrasound in an aqueous solution are responsible for the oxidation reaction. Luche [69] investigated sonochemical reactions occurring in a heterogeneous system. Jana and Chatterjee [70] made an estimation of hydroxyl free radicals produced by ultrasound in Fricke solution using Fricke dosimeter. The dose–response relation was found to be linear for different intensities of ultrasound. 20 kHz frequency ultrasound produces 14 times more hydroxyl radicals than those produced by 3.5 MHz. Henglein [71] stated that the free radicals produced by the cavitation effect are responsible for reaction. The OH radicals produced by the ultrasound are strong oxidation agents and lead to H_2O_2 formation.

Hoffmann et al. [72] investigated sonochemical degradation of organic compounds present in water. Three distinct pathways of sonochemical degradation of organic compounds by acoustic cavitation have been proposed: (1) Oxidation by hydroxyl radicals, (2) Pyrolytic decomposition, and (3) Supercritical water oxidation. Gogate et al. [73] analyzed and mapped sonochemical reactors by experimental verification. Generalized correlations were developed for effective design and scale-up of sonochemical reactors. Decomposition of potassium iodide was taken as a model reaction; iodine liberation from the reaction is only by cavitation effect, and not by shear temperature and pressures. This is because free OH^- radicals are formed in the solution only under cavitating conditions. Controlling reduced sulfur compounds by using hydrogen peroxide has been investigated and reported. Hydrogen peroxide combines advantages not obtainable with any other single form of chemical control. It is cost-effective and specific, forming no toxic by-products. It is safe to work with when handled properly, and produces soluble sulphates, thus avoiding the sludge problem. Hydrogen peroxide has been used for industrial purposes for a long time because of its physical and chemical nature, i.e., low freezing point, unlimited solubility in water, and reactivity.

Given the wealth of data available in literature pertaining to formation of free radicals in water irradiated with ultrasonics, no specific attempt was made as part of this study to confirm or quantify the presence of such species. Instead, this study focuses on the physical aspects of ultrasonic desulfurization, namely cavitation, streaming and their combined effects. Since the ultrasonic systems used in this study are state-of-the-art with respect to energy transmission and uniformity characteristics, it was felt that the emphasis in the study should be placed on investigating the effects of ultrasonic field parameters such as frequency and amplitude.

2.5 Ultrasonic Process for De-Ashing and De-Sulfurization

Very few researchers have focused on ultrasonic coal de-sulfurization, Existing literature fails to explain the mechanisms involved in ultrasound-assisted coal de-sulfurization. Conclusions drawn by researchers are very general in nature. The ultrasonic de-sulfurization methods studied are either aqueous or chemical based. The biggest advantage of ultrasonic method is simultaneous removal of ash and sulfur. Zaidi [74] investigated ultrasound-promoted de-sulfurization of low-rank coals with dilute solutions of sodium hydroxide (0.025 to 0.2 M) at 30 and 70 °C. The sulfur removal was higher for samples sonicated at a lower temperature. The shear forces produced by the ultrasound energy are responsible for exposing the finely disseminated sulfur sites in coal to alkali attack. However, the mechanism involved in the interaction between sonication and dilute sodium hydroxide is not explained. Ze et al. [75] investigated the enhancement of de-sulfurization and de-ashing of coal. 100 g of Zibo coal and 300 ml of water mixture were sonicated for 10 min using 20 kHz frequency and 200 W power. Then, the sample was wet screened. The same procedure was followed without sonication. Yield, sulfur and ash analysis were performed, and results revealed that ultrasonic conditioning can drive physical separation of pyrite and refuse from coal. On the other hand, ultrasonic conditioning can change the surface of the coal and pyrite particles, and increase the hydrophobicity of slime and the hydrophilicity of pyrite and refuse. For a 1.3 kg/t flotation agent-to-coal ratio, the perfect index of flotation, the perfect index of desulfurization and the percentage of desulfurization after ultrasonic processing increased by 22.51, 25.36 and 2.49%, respectively. It may be concluded that ultrasonic conditioning can, in general, enhance the performance of coal flotation methods used for desulfurization and de-ashing.

Grobas et al. [76] investigated hydrogenation of cyclohexene, biphenyl, and quinoline, and hydro-desulfurization of benzothiophene in the presence of formic acid (a hydrogen precursor), and a Pd/C catalyst; ultrasound irradiation was investigated as well. It was found that the use of formic acid in the presence of ultrasonic irradiation was effective in promoting hydrogenation and desulfurization at very mild conditions (i.e., ambient temperature and pressure).

Wang et al. [77] used several carbon-based sorbents for de-sulfurization of a model jet fuel. The results showed that the selective adsorption ability of $PdCl_2$ was higher than those of CuCl and metallic Pd. The results of desorption experiments showed that ultrasound-assisted regeneration was an effective method for the saturated $PdCl_2$/AC that was saturated with benzothiophene and substituted compounds. The amount of sulfur desorbed was higher with ultrasound, 65 wt% desorption vs. 45 wt% without ultrasound.

Mello et al. [78] investigated ultrasound-assisted oxidative process for sulfur removal from petroleum product feedstock. Dibenzothiophene is used as a model sulfur compound. The effect of sonication time, volume of oxidizing reagents, kind of solvent for the extraction step and kind of organic acid were investigated. Higher efficiency of sulfur removal was achieved using sonication in comparison to experiments performed without its application, under the same reaction conditions.

At present, Coal India Limited (CIL) operates 17 coal washeries, out of which 11 are for coking coal and the remaining are for non-coking coal, with a total capacity of 39.40 million tonnes per annum. CIL has been operating coking as well as non-coking coal washeries for a long period of time, but due to manpower constraints and operational cost, it has decided to outsource coal washery operation to private players. A number of private operators are already involved in coal washing in the country, and the power sector has started using washed coal for power generation, considering its economic and environmental benefits. Bilaspur Coal Washery produces 3 million tones per annum. Aryan Coal Washery produces nearly 22 million tones of coal per year. Tata Steel and SAIL's joint-venture will set up a 1.8 million tonnes per annum (mtpa) coal washery at Bhelatand in Jharkhand, through an investment of Rs 200 crore. Twenty new washeries with an annual capacity of 111 million tonnes per annum are being taken up by Coal India (CIL) during the XI and XII Five-Year Plans.

There are several references in technical and trade literature regarding the employment of ultrasonics in coal cleaning and beneficiation; it is clearly a highly-scalable process that is in widespread use globally. Intuitively, there is no reason to question the wisdom of its implementation for Indian coals, whose large ash content renders them very suitable for this purpose. This study attempts to validate ultrasonic wash for Indian coals on a laboratory scale, with the ultimate objective of defining a scalable process that can be implemented in production quantities.

2.6 Scope and Objectives for the Present Work

There are a number of methods available for remediation of ash related problems, but economically all have disadvantages. The utilization of ultrasound for the treatment of waste materials is a growing area of sonochemical research. Power ultrasound can be used for the beneficiation of coal by the removal of mineral matter (ash and sulfur) from coal. The following is a summary of key observations from a study of related literature:

- In conventional coal wash, the main focus is on surface cleaning, but there is little focus on interior part of the coal matrix.
- Ultrasonic coal wash method involves numerous influencing factors, and several complex, interdependent mechanisms have not been studied.
- Interaction mechanism between suspended coal particle and ultrasound has not been fully investigated.
- No attempt has been made to understand the mechanism of coal particle breakage and ash removal as a function of ultrasonic field parameters.
- So far, sono-fragmentation research has been primarily based on probe-type sonicator, not tank-type.

Ultrasonic coal-wash is thus widely used outside India, but not well understood by researchers. Hence, there is a necessity for experimental investigation and modeling to characterize the complex mechanisms and the effect of influencing parameters.

The specific objectives of present work are

1. Assessment of Fly–Ash Erosion Potential of Indian Coals
2. Experimental Studies on Ultrasonic Coal Beneficiation
   (a) Ultrasonic Aqueous-Based Coal Beneficiation
      i. De-Ashing
      ii. De-Sulfurization
   (b) Ultrasonic Reagent-Based Coal Beneficiation
      i. De-Ashing
      ii. De-Sulfurization
3. Experimental Optimization and Mechanistic Modeling of Ultrasound Assisted Reagent-Based Coal De-Sulfurization
4. Assessment of Benefits from Ultrasonic Coal-Wash (USCW)

References

1. Finnie I (1960) Erosion of surfaces by solid particles. Wear 3:87–103
2. Hutchings IM, Winter RE (1974) Particle erosion of ductile materials: a mechanism of material removal. Wear 27:121–128
3. Jennings WH, Head WJ Jr, Mannings CR (1976) A mechanistic model for the prediction of ductile erosion. Wear 40:93
4. Soo SL (1977) A note on erosion by moving dust particles. Powder Technol 17:259–263
5. Foley T, Levy A (1983) The erosion of heat-treated steels. Wear 91:45–64
6. Sundararajan G, Shewmon PG (1983) A new model for the erosion of metals at normal incidence. Wear 84:237–258
7. Levy AV, Yan J, Patterson J (1986) Elevated temperature erosion of steels. Wear 108:43–60
8. Meng HC, Ludema KC (1995) Wear models and predictive equations: their form and content. Wear 181:443–457
9. Wang BQ (1995) Erosion-corrosion of coatings by bio-mass-fired boiler fly ash. Wear 188:40–48
10. Xie J, Walsh PM (1995) Erosion-corrosion of carbon steel by products of coal combustion. Wear 186:256–265
11. Hubner W, Leitel E (1996) Peculiarities of erosion-corrosion processes. Tribol Int 29:199–206
12. Oka YI, Olmogi H, Hosokawa T, Matsumura M (1997) The impact angle dependence of erosion damage caused by solid particle impact. Wear 203:573–579
13. Hussainova I, Kubarsepp J, Shcheglov I (1999) Investigation of impact of solid particles against hard metal and cermets targets. Tribol Int 32:337–344
14. Lyczkowski RN, Bouillard JX (2002) State-of-the-art review of erosion modeling in fluid/solids systems. Prog Energy Combust Sci 28:543–602
15. Mbabazi JG, Sheer TJ, Shandu R (2004) A model to predict erosion on mild steel surfaces impacted by boiler fly ash particles. Wear 257:612–624
16. Marcus K, Moumakwa DO (2005) Tribology in coal-fired power plants. Tribol Int 38:805–811
17. Das SK, Godiwalla KM, Mehrotra SP, Sastry KKM, Dey PK (2006) Analytical model for erosion behavior of impacted fly-ash particles on coal-fired boiler components. Sadhana 31:583–595
18. Vicenzi J, Villanova DL, Lima MD, Takimi AS, Marques CM, Bergmann CP (2006) HVOF-coatings against high temperature erosion (3,000 C) by coal fly ash in thermoelectric power plant. Mater Des 27:236–242
19. Wang YF, Yang ZG (2008) Finite element model of erosive wear on ductile and brittle materials. Wear 265:871–878
20. Steinberg M, Yang RT, Horn TK, Berlad AL (1977) Desulfurization of coal with ozone: an attempt. Fuel 56:227–228
21. Araya PE, Ohlbaum RB, Droguett SE (1981) Study of the treatment of sub bituminous coals by NaOH solutions. Fuel 60:1127–1130
22. Chandra D, Chakrabarti JN, Swamy YV (1982) Auto-desulfurization of coal. Fuel 61:204–205
23. Krzymien LL (1982) Complete removal of sulfur from coal using solutions containing cupric ions. Fuel 61:871–873
24. Chaung KC, Markuszewesky R, Wheelock TD (1983) Desulfurization of coal by oxidation in alkaline solutions. Fuel Process Technol 7:43–57
25. Yang RT, Das SK, Tsai BMC (1985) Coal demineralization using sodium hydroxide and acid solutions. Fuel 65:735–742
26. Kara H, Ceylan R (1988) Removal of sulfur from four central anatolian lignite's by NaOH. Fuel 67:170–172

27. Ahnonkitpanit E, Prasassarakich P (1989) Coal desulfurization in aqueous hydrogen peroxide. Fuel 68:819–824
28. Ozdemmir M, Bayrakceken S, Gurses A, Gulaboglu S (1990) Desulfurization of two turkish lignite's by chlorinolysis. Fuel Process Technol 26:15–23
29. Ali A, Srivatsava SK, Haque R (1992) Chemical desulfurization of high sulfur coals. Fuel 71:835–839
30. Prasassarakich P, Thaweesri T (1996) Kinetics of coal desulfurization with sodium benzoxide. Fuel 75:816–820
31. Rodriguez RA, Jul CC, Limon DG (1996) The influence of process parameters on coal desulfurization by nitric leaching. Fuel 75:606–612
32. Hamamci C, Kahraman F, Diiz MZ (1997) Desulfurization of southeastern anatolian asphaltites by the Meyers method. Fuel Process Technol 50:171–177
33. Aacharya C, Kar RN, Sukla LB (2001) Bacterial removal of sulfur from three different coals. Fuel 80:2207–2216
34. Mukherjee S, Borthakur PC (2001) Chemical demineralization/desulfurization of high sulfur coal using sodium hydroxide and acid solutions. Fuel 80:2037–2040
35. Ratanakandilok S, Ngamprasertsith S, Prasassarakich P (2001) Coal desulfurization with methanol/water and methanol/KOH. Fuel 80:1937–1942
36. Sonmez O, Giray ES (2001) The influence of process parameters on desulfurization of two turkish lignite's by selective oxidation. Fuel Process Technol 70:159–169
37. Aacharya C, Sukla LB, Misra VN (2005) Biological elimination of sulfur from high sulfur coal by aspergillus-like fungi. Fuel 84:1597–1600
38. Baruah BP, Saikia BK, Kotoky P, Rao PG (2006) Aqueous leaching on high sulfur sub-bituminous coals, in Assam, India. Energy Fuels 20:1550–1555
39. Liu K, Yang JJ, Wang Y (2008) Desulfurization of coal via low temperature atmospheric alkaline oxidation. Chemosphere 71:183–188
40. Yuda Y, Ayse T (1990) Supercritical extraction and desulfurization of Beypazari lignite by ethyl alcohol/NaOH treatment. 1. Effect of ethyl alcohol/coal ratio and NaOH. Pet Sci Technol 8:87–105
41. Li W, Guo S (1996) Supercritical desulfurization of high rank coal with alcohol/water and alcohol/KOH. Fuel Process Technol 46:143–155
42. Charutawai K, Ngamprasertsisith S, Prasassarakich P (2003) Supercritical desulfurization of low rank coal with ethanol/KOH. Fuel Process Technol 84:207–216
43. Mukherjee S, Borthakur PC (2003) Effect of leaching high sulfur sub bituminous coal by potassium hydroxide and acid on removal of mineral matter and sulfur. Fuel 82:783–788
44. Mukherjee S, Borthakur PC (2004) Demineralization of sub bituminous high sulfur coal using mineral acids. Fuel Process Technol 85:157–164
45. Alam HG, Mogaddam AG, Omidkhah MR (2009) The influence of process parameters on desulfurization of Mezino coal by HNO_3/HCl leaching. Fuel Process Technol 90:1–7
46. Fridman VM (1972) The interaction mechanism between cavitation bubbles and particles of the solid and liquid phases. Ultrasonics 10:162–165
47. Kusters KA, Pratsinis SE, Thoma SG, Smith DM (1993) Ultrasonic fragmentation of agglomerate powders. Chem Eng Sci 48:4119–4127
48. Kusters KA, Pratsinis SE, Thorna SG, Smith DM (1994) Energy-size reduction laws for ultrasonic fragmentation. Powder Technol 80:253–263
49. Gopi KR, Nagarajan R (2008) Application of power ultrasound in cavitation erosion testing of nano-ceramic particle/polymer composites. Solid State Phenom 136:191–204
50. Raman V, Abbas A (2008) Experimental investigations on ultrasound mediated particle breakage. Ultrason Sonochem 15:55–64
51. Newman AP, Lorimer JP, Mason TJ, Hutt KR (1997) An investigation into the ultrasonic treatment of polluted solids. Ultrason Sonochem 4:153
52. Kruger O, Schulze TL, Peters D (1999) Sonochemical treatment of natural ground water at different high frequencies. Ultrason Sonochem 6:123–128

53. Farmer AD, Collings AF, Jameson GJ (2000) The application of power ultrasound to the surface cleaning of silica and heavy mineral sands. Ultrason Sonochem 7:243–247
54. Kim YU, Wang MC (2003) Effect of ultrasound on oil removal from soils. Ultrasonics 41:539–542
55. Mason TJ, Collings A, Sumel A (2004) Sonic and ultrasonic removal of chemical contaminants from soil in the laboratory and on a large scale. Ultrason Sonochem 11:205–210
56. Cooke NE, Fuller OM, Gaikwad RP (1989) Ultrasonic extraction of coal. Fuel 68:1227–1233
57. US Patent # 4741839, European Patent EP0259959 (1988) Ultrasonic vibrator tray processes and apparatus
58. British Patent # GB 2,139,245 (1984) ~ (CA 102:64815) ~ Coal cleaning with ultrasound
59. US Patent # 4,156,593 (1979) ~ (CA 91:94260) ~ "Ultrasonic wet-grinding coal"
60. Lindstrom O, Lamm O (1951) The chemical effects produced by ultrasonic waves. J Phys Chem 55:1139–1146
61. Webster E (1963) Cavitation. Ultrasonics 1:39–48
62. Makino K, Mossoba MM, Riesz P (1982) Chemical effects of ultrasound on aqueous solutions evidence for OH^- and H^+ by spin trapping. J Am Chem Soc 104:3537–3539
63. Riesz P, Berdahl D, Christman CL (1985) Free radical generation by ultrasound in aqueous and non-aqueous solutions. Environ Health Perspect 64:233–252
64. Christman CL, Carmichael AJ, Mossoba MM, Riesz P (1987) Evidence for free radicals produced in aqueous solutions by diagnostic ultrasound. Ultrasonics 25:31–34
65. Misik V, Riesz P (1994) Free radicals formation by ultrasound in organic liquids: a spin trap and EPR study. J Phys Chem 98:1634–1640
66. Misik V, Riesz P (1996) Peroxyl radical formation in aqueous solutions of n-dimethylformamide, n-methylformamide, and dimethylsulfoxide by ultrasound. Free Radical Biol Med 20:129–138
67. Margulis MA (1994) Fundamental problems of sonochemistry and cavitation. Ultrason Sonochem 1:87–90
68. Entezari MH, Krus P (1994) Effect of frequency on sono chemical reaction. Ultrason Sonochem 1:75–79
69. Luche JL (1994) Effect of ultrasound on heterogeneous system. Ultrason Sonochem 1: 111–118
70. Jana AK, Chatterjee SN (1995) Estimation of hydroxyl free radicals produced by ultrasound in Fricke solution used as a chemical dosimeter. Ultrason Sonochem 2:87–91
71. Henglein A (1995) Chemical effects of continuous and pulsed ultrasound in aqueous solutions. Ultrason Sonochem 2:115–121
72. Hoffmann MR, Hua I, Hochemer R (1996) Application of ultrasonic irradiation for the degradation of chemical contaminants in water. Ultrasonic Sonochemistry 3:163–172
73. Gogate PR, Tatake PA, Kanthale PM, Pandit AB (2002) Mapping of sonochemical reactors; Review, analysis and experimental verification. AIChE J 48:1542–1560
74. Zaidi SAH (1993) Ultrasonically enhanced coal desulfurization. Fuel Process Technol 33: 95–100
75. Ze KW, Xin XH, Tao CJ (2007) Study of enhanced fine coal desulfurization and de-ashing by ultrasonic floatation. J Chin Univ Min Technol 17:358–362
76. Grobas J, Bolivar C, Scott CE (2007) Hydro-desulfurization of benzothiophene and hydrogenation of cyclohexene, biphenyl, and quinoline, assisted by ultrasound, using formic acid as hydrogen precursor. Energy Fuels 21:19–22
77. Wang Y, Yang RT (2007) Desulfurization of liquid fuels by adsorption on carbon-based sorbents and ultrasound-assisted sorbent regeneration. Langmuir 23:3825–3831
78. Mello PA, Duarte FA, Nunes MAG, Alencar MS, Moreira EM, Korn M, Dressler VL, Flores ÉMM (2009) Ultrasound-assisted oxidative process for sulfur removal from petroleum product feedstock. Ultrason Sonochem 16:732–736

Chapter 3
Assessment of Fly-Ash Erosion Potential of Indian Coals

3.1 Introduction

Erosion is defined as a process by which material is removed from the layers of a surface impacted by a stream of abrasive particles. Fly-ash particles entrained in the flue gas from boiler furnaces in coal-fired power plants can cause serious erosive wear on steel surfaces along the flow path, thereby reducing the operational life of the mild-steel heat transfer plates that are used in the rotary regenerative heat exchangers. Moreover, in technical practice, erosion is often accompanied by a chemical attack. The variables which influence the wear process can be separated and studied independently. In the present study, the effect of ash-particle physical properties and transport dynamics on the erosive wear of three different grades of low alloy steel (old and new coupon), using three different power-station ash types, is investigated. The main difference between old and new coupon is surface roughness. The study used a Taguchi fractional-factorial L_{27} DOE. Taguchi's parameter design is a general method for design which provides a simple and systematic approach for optimization of design for quality, performance and cost. The selection of control factors is the most important stage in the experimental design. Taguchi creates a standard orthogonal array to address this requirement. Orthogonal experiment enables to investigate the relative importance of various factors and identify the best levels for different factors on a response and the results can be analyzed by using a common mathematical procedure. This method can significantly reduce experimental time and research cost by using orthogonal arrays. The number of trials chosen for an experimental design is based on the resolution desired and the number of the chosen experimental parameter levels is based on the range of operating conditions of the process [1]. It is this aspect of the Taguchi's fractional-factorial method that is used in the present investigation to set the parameter values for data collection. The experimental data thus obtained are then used to derive a model for prediction of erosion. The model incorporates the properties of the ash particles and the target metal surface, as well

B. Ambedkar, *Ultrasonic Coal-Wash for De-Ashing and De-Sulfurization*, Springer Theses, DOI: 10.1007/978-3-642-25017-0_3,

Table 3.1 Properties of low alloy steel

Coupon type	Microstructure (ASTM E407)	ASTM ferrite grain size	Mn (%)	Si (%)	C (%)	Cr (%)	Mo (%)
Grade 12	Polygonal grains of ferrite and tempered bainite	Finer than 8	0.5	0.20	0.08	0.85	0.44
Grade 22	Fine tempered bainite	N/A	0.5	0.10	0.12	2.38	0.89
Grade A	Polygonal grains of ferrite and pearlite	Finer than 8	0.79	0.12	0.12	<0.1	<0.1

as the characteristics of ash particle motion in the form of the impingement velocity and the impingement angle.

3.2 Experimental Details

3.2.1 Materials Used and Their Properties

Table 3.1 lists the physical properties and chemical compositions of the various grades of low-alloy steels used in the study. Table 3.1 shows that % of Mn, Si, C, Cr and Mo present in the various grades of low alloy steel; rest of the element is Iron. The % of chromium in grade 22 low-alloy steel is higher than that of other two low-alloy steel, remaining elements show limited variation.

Size distribution analysis has been made on three different thermal power plant fly-ash (Mettur, Raichur and Dhahanu thermal power plant) sample using sieving analysis (dry method), because particle size and shape plays the major role in erosion phenomena. Particle size distributions of the three ash-types tested are shown in Fig. 3.1(a–c). The feed size of coal for coal burning boilers is less than 75 microns. The size measured seems to be bigger than the initial coal feed size due to agglomeration of fly-ash particle at high temperature. Figure 3.1a shows the size distribution analysis of Mettur fly ash sample, where more than 75% of the fly ash particle size come under $(-150 + 100)$ and $(-250 + 150)$ microns range. In the case of Raichur thermal power plant fly ash, the distribution of particle size is fairly uniform throughout the fly ash sample, but in Dhahanu thermal power plant fly ash sample, there is a narrow distribution of size, i.e., more than 80% of the particle belonging to $(-250 + 210)$ microns size. This may be due to nature of coal and process conditions of that particular thermal power plant. In each ash type, three different size ranges of fly-ash particle $(-75 + 0$, $-150 + 75$ and $-250 + 150$ micron) were chosen to determine the effect of fly-ash particle size on erosion.

Table 3.2 provides the fly-ash chemical compositions. From Table 3.2, it has been observed clearly that, fly-ash mainly consists more than 90% of silica and

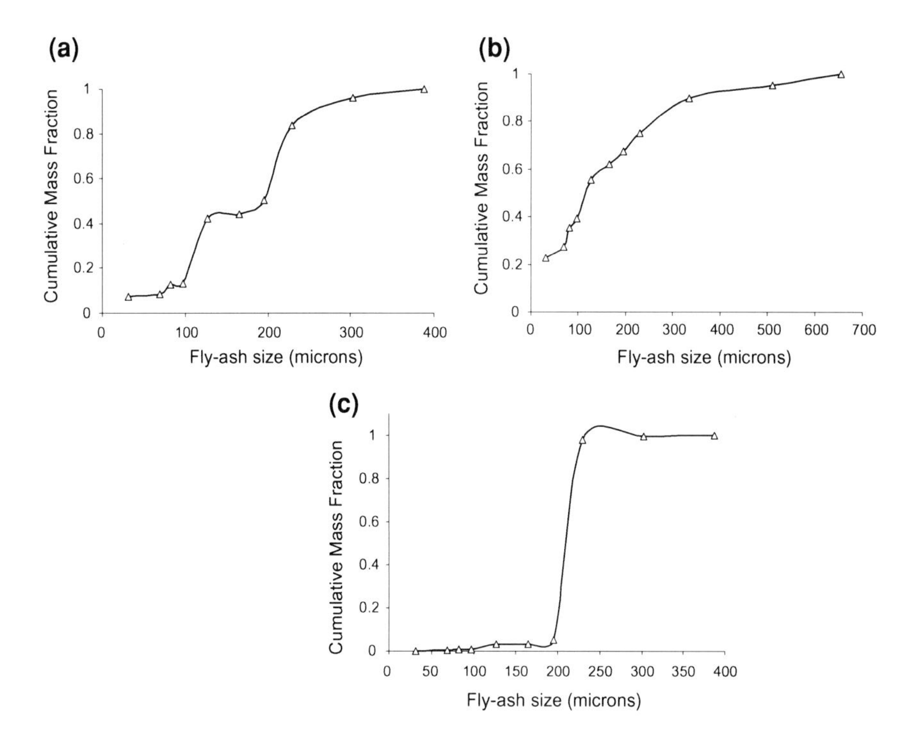

Fig. 3.1 **a** Size distribution analysis of Mettur fly-ash. **b** Size distribution analysis of Raichur fly-ash. **c** Size distribution analysis of Dhahanu fly-ash

Table 3.2 Ash properties (wt%)

Source	SiO_2	Al_2O_3	TiO_2	Fe_2O_3
Mettur	62.2	28.3	0.9	6.0
Raichur	60.2	25.5	1.8	6.0
Dhahanu	64.8	25.1	2.2	5.3

Table 3.3 Coupon surface roughness and hardness prior to start of erosion test

Coupon type	Mean hardness, HV_{10}	Ra, μm (Old coupon)	Ra, μm (new coupon)
Grade 12 steel	178	2.01	1.7142
Grade 22 steel	204	3.084	1.6571
Grade A steel	130	3.45	1.0985

alumina. Table 3.3 shows surface roughness and hardness values of coupons prior to start of erosion test. Average surface roughness value of target material is measured all over sample surface using Profilometer. Main difference between aged (old) and new coupon is surface roughness and is shown in Table 3.3. Hardness of grade 22 low-alloy steel is higher than the other two.

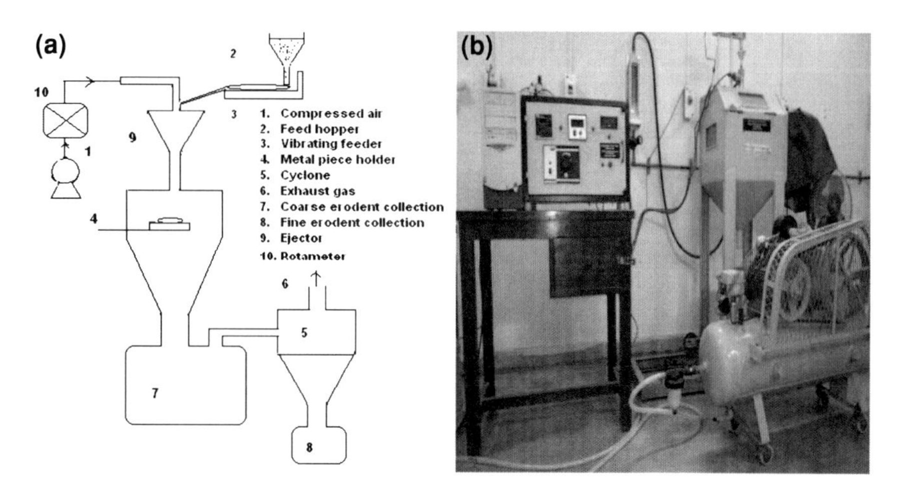

Fig. 3.2 **a** Schematic of Air-Jet erosion tester. **b** Photographic view of Air-Jet erosion tester

3.2.2 *Test Equipment and Procedure*

3.2.2.1 Experimental Set Up

A Sand-blast type air-jet erosion test rig, shown schematically in Fig. 3.2a and photographic view in Fig. 3.2b, was used in this study to carry out airborne particulate erosion tests. The erodent particles were entrained in a stream of compressed air and accelerated down a 65 mm long nozzle with 4.55 mm internal diameter; they were then made to impact on a target mounted on angle fixtures (15, 30 and 45°). The target material is fixed 10 mm away from the nozzle. Compressed air is sent from air compressor through Rota-meter, where the velocity of air is measured. The fly-ash is supplied by vibrating feeder through nozzle. The slip velocity between air and fly-ash is negligible because compressed air and the erodent are passed through the same nozzle and the erodent attains velocity of air prior to impaction. Also, the distance between the nozzle and the target material is very small (10 mm).

Based on the magnitude of vibration the fly-ash feed rate is fixed (1 g/min). The target material can be fixed on the metal piece holder, where the angle of the target material can also be changed by adjusting screws on the metal holder. The tested fly-ash particle and eroded materials are collected in the bottom of the erosion test chamber; cyclone is connected to this chamber to separate fine erodent from coarser one.

3.2.2.2 Experimental Procedure

Airborne erosion tests were carried out for three types of metal target. The specimen's size is (40 × 40 × 5 mm) cut-pieces. The initial weight of the target was measured using an electronic micro-balance. The target specimens were

Table 3.4 Taguchi L_{27} DOE: parameters and levels for erosion experiments

Factor	Level (3)
Velocity	10, 20, 30 (m/s)
Angle	15, 30, 45 (degrees)
Time	5, 10, 15 (min)
Feed quantity	5, 10, 15 (g @ 1 g/min feed-rate)
Particle size	(−75 + 0), (−150 + 75), (−250 + 150) μm
Ash type	Mettur, Raichur, Dhahanu

Table 3.5 L_{27} orthogonal array for erosion experiments

Trial no.	Velocity, m/s	Angle, degree	Size, μm	Ash type	Feed qty, g	Time, min
1	10	15	(−75 + 0)	Mettur	5	5
2	10	15	(−150 + 75)	Raichur	10	10
3	10	15	(−250 + 150)	Dhahanu	15	15
4	10	30	(−75 + 0)	Mettur	10	10
5	10	30	(−150 + 75)	Raichur	15	15
6	10	30	(−250 + 150)	Dhahanu	5	5
7	10	45	(−75 + 0)	Mettur	15	15
8	10	45	(−150 + 75)	Raichur	5	5
9	10	45	(−250 + 150)	Dhahanu	10	10
10	20	15	(−75 + 0)	Raichur	5	5
11	20	15	(−150 + 75)	Dhahanu	10	10
12	20	15	(−250 + 150)	Mettur	15	15
13	20	30	(−75 + 0)	Raichur	10	10
14	20	30	(−150 + 75)	Dhahanu	15	15
15	20	30	(−250 + 150)	Mettur	5	5
16	20	45	(−75 + 0)	Raichur	15	15
17	20	45	(−150 + 75)	Dhahanu	5	5
18	20	45	(−250 + 150)	Mettur	10	10
19	30	15	(−75 + 0)	Dhahanu	5	5
20	30	15	(−150 + 75)	Mettur	10	10
21	30	15	(−250 + 150)	Raichur	15	15
22	30	30	(−75 + 0)	Dhahanu	10	10
23	30	30	(−150 + 75)	Mettur	15	15
24	30	30	(−250 + 150)	Raichur	5	5
25	30	45	(−75 + 0)	Dhahanu	15	15
26	30	45	(−150 + 75)	Mettur	5	5
27	30	45	(−250 + 150)	Raichur	10	10

mounted 10 mm from nozzle orifice for all impingement angles. Table 3.4 lists the controlled parameters and their level settings employed in the Taguchi DOE. Table 3.5 shows L_{27} DOE. The experimental conditions were chosen as per the trial conditions given in Table 3.5. At the end of the each test, the target sample was taken out, cleaned and weighed to calculate weight loss. The resolution of the balance used is 0.01 mg. This is adequate for the measurements performed. This weight loss was taken to represent erosion mass loss.

3.3 Results and Discussion

In our experiment, a total 162 erosion data points were measured and is given in Appendix I. Each trial was conducted twice and the averaged value is tabulated in the appendix section. It should be pointed out that following traditional Taguchi DOE methodology, effects of all other controllable parameters have been averaged over the entire DOE; this enables isolation of effects due to a single variable. This methodology is followed in Figs. 3.3, 3.4, 3.5, 3.6 and 3.7 as well.

3.3.1 Effect of Process Parameters on Erosion Mass Loss

Figures 3.3a, b show the measured effect of fly-ash particle impact velocity on room-temperature metal erosion. The erosion rate increases with increasing fly-ash particle impact velocity on metal surfaces. The erosion mass loss is highest for Grade A (aged coupon) (Fig. 3.3a) due to high surface roughness. Similarly, from Fig. 3.3b lowest erosion mass loss was observed for Grade A (new coupon) due to lowest surface roughness value. The surface roughness value of three different low alloy steel is shown in Fig. 3.3c. In Fig. 3.4, the measured effect of impact angle is shown. Erosion rate generally decreases with an increase in the impingement angle, a behavior consistent with classical ductile erosion [2–4].

Effect of impacting particle size on erosion is shown in Fig. 3.5. Experiments were carried out with three different average sizes of the fly ash particles (37.5, 112.5 and 200 μm). Erosion rate is observed to increase with an increase in particle size up to about 112.5 μm, and to level-off thereafter. Fly-ash tested in this size range of $(-150 + 75)$ μm comprises a greater amount of silica, leading to more pronounced impaction due to high velocity compared to the other two. Effect of fly-ash mass loading on erosive wear is indicated in Fig. 3.6; as ash particle loading increases, so does loss of material by erosion. From Fig. 3.7, it is evident that erosion proceeds monotonically with time of exposure, at least for short durations of exposure.

3.3.2 Effect of Surface Roughness on Erosion Mass Loss

Grade 22 low-alloy steel was taken as a target material, 112.5 micron average size Mettur fly-ash particle was chosen as erodent, 30° impact angle and 20 m/s impact

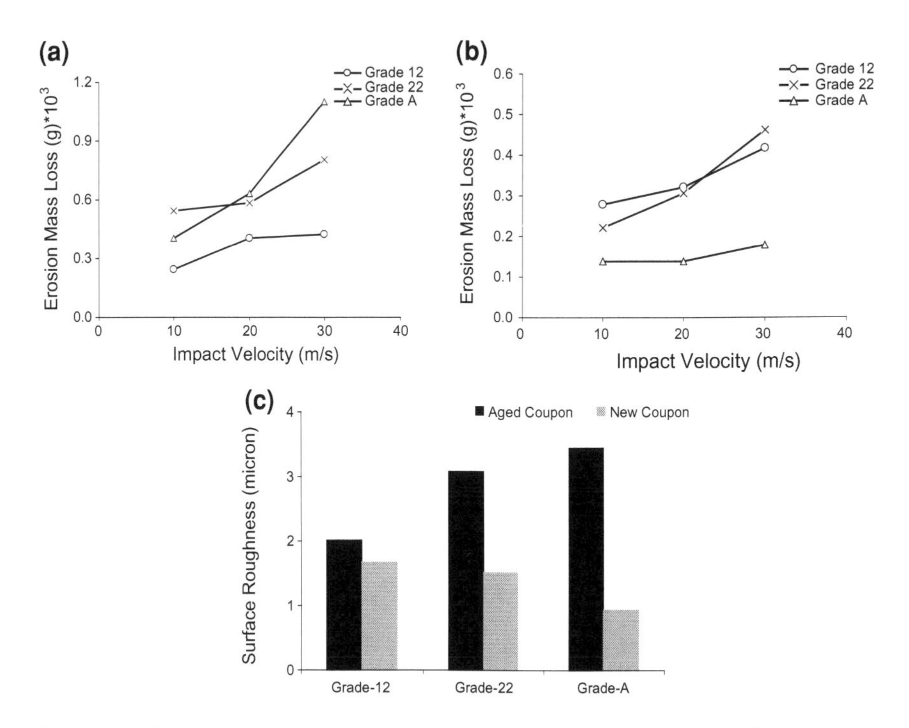

Fig. 3.3 **a** Effect of impact velocity on erosion mass loss for aged coupon. **b** Effect of impact velocity on erosion mass loss for new coupon **c** Surface roughness values of various grade of low-alloy steel

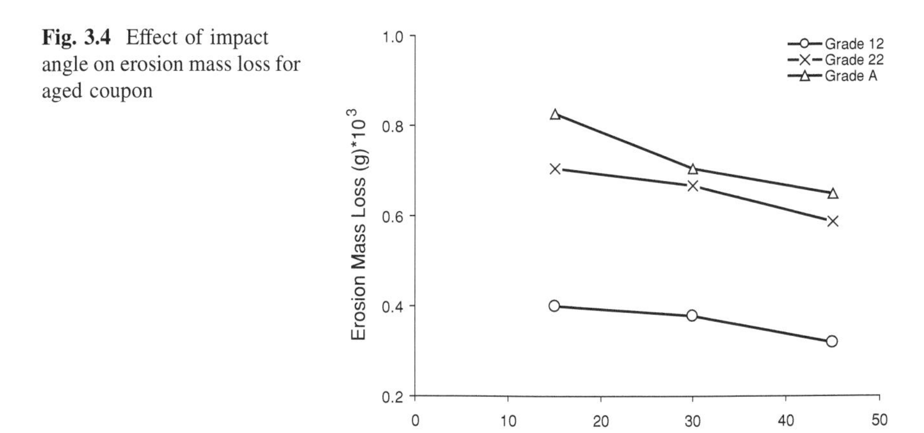

Fig. 3.4 Effect of impact angle on erosion mass loss for aged coupon

velocity were used as process conditions. Some point on the target material was marked as a target point, initial surface roughness of that point was measured using profile-meter and the initial weight of coupon was also measured using

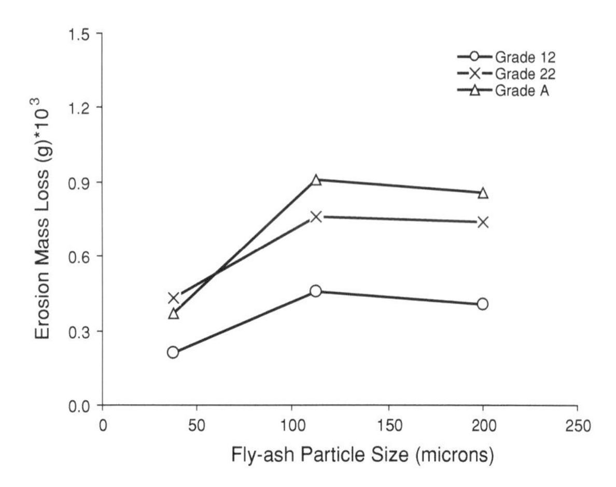

Fig. 3.5 Effect of ash particle size on erosion mass loss for aged coupon

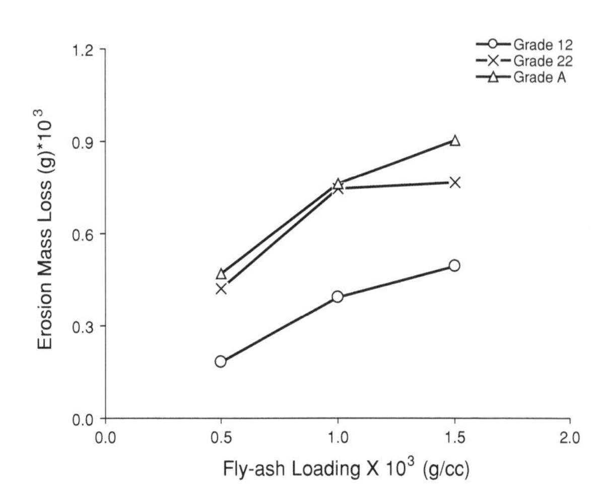

Fig. 3.6 Effect of fly-ash loading on erosion mass loss for aged coupon

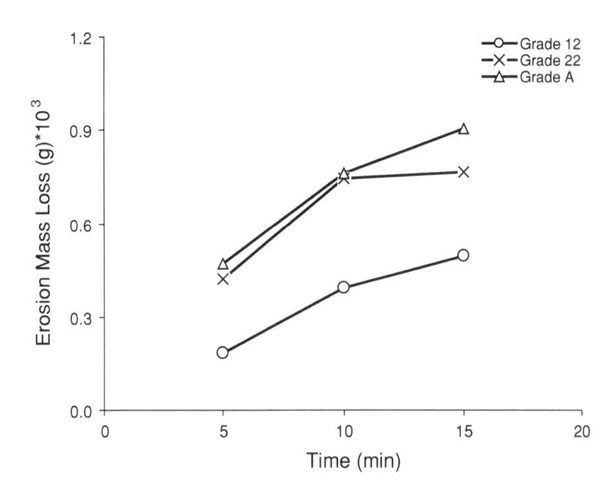

Fig. 3.7 Effect of exposure time on erosion mass loss for aged coupon

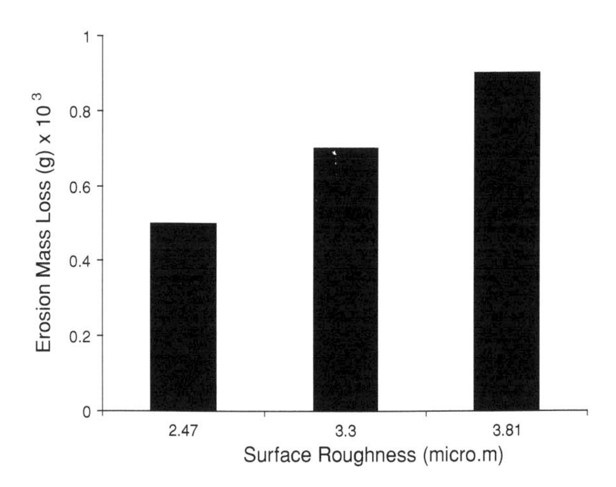

Fig. 3.8 Effect of surface roughness on erosion mass loss

micro-balance. Erosion test was conducted for 10 min the erosion mass loss is plotted against surface roughness of low-alloy steel and is shown in Fig. 3.8. Erosion mass loss is larger for high surface roughness low-alloy steel was observed.

3.3.3 Effect of Ash and Substrate Alloy Properties on Erosion Mass Loss

The effect of fly-ash properties (chemical-compositional) on erosion potential is displayed in Fig. 3.9. As titania-content increase, there is an associated reduction in the erosive effect due to ash impaction; the same trend is observed for all substrate materials tested. Interestingly, bulk-material properties of the substrate alloys; such as composition and hardness (Fig. 3.10a, b) have a secondary effect on erosion susceptibility compared to surface characteristics, such as roughness (Fig. 3.10a). This would lead one to conclude that incipient and even short-life room-temperature erosion is primarily dependant on surface characteristics.

3.3.4 Predictive Model for Ash-Impact Erosion

Given the availability of large number of experimental data, it becomes possible to predict the relationship between erosion loss and the factors that are influencing the rate of room-temperature erosion. Least Square Method was employed to develop this erosion model. General form of erosion equation is given in Eq. 3.1. The experimental data's, ash and substrate alloy properties are used to develop this empirical model.

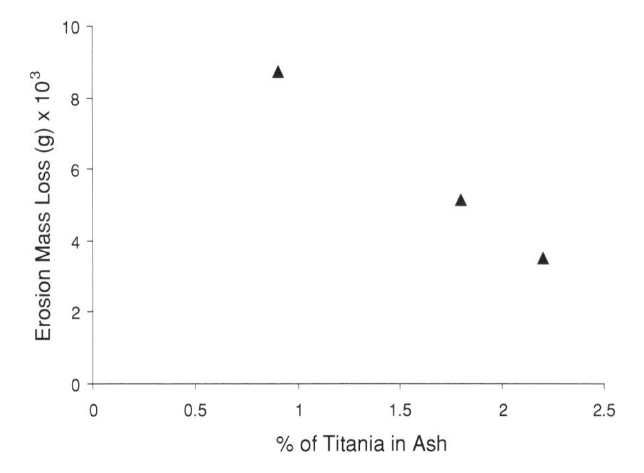

Fig. 3.9 Effect of ash properties on erosion mass loss

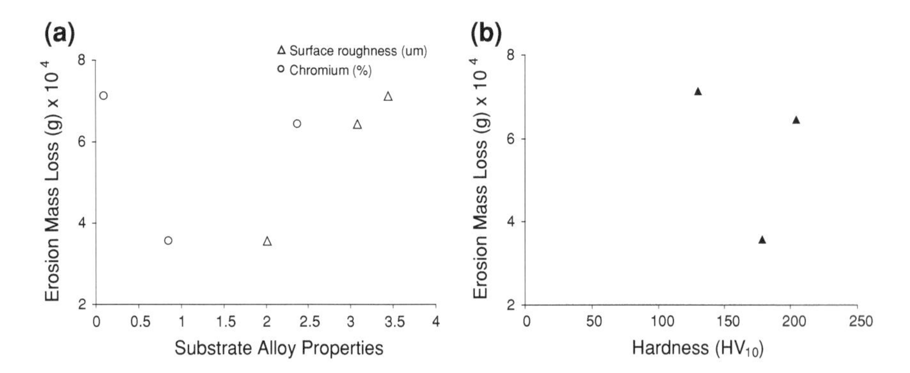

Fig. 3.10 **a** Effect of substrate alloy properties on erosion mass loss for aged coupon. **b** Effect of hardness of low-alloy steel on erosion mass loss for old coupon

$$E = KV_P^a d_P^b \left(\frac{Q_P}{V_F}\right)^C (Sin\alpha)^d (R_a)^e M^f (Ti)^g (Cr)^h \tag{3.1}$$

Where,

E = Erosion loss (g)
K = Empirical Constant
V_p = Fly ash particle impact velocity (m/s)
d_p = Fly-ash particle size (microns)
Q_p = Particle quantity (kg/min) (5, 10, 15 g/min)
V_F = Volumetric flow rate (m^3/min) [Area of Nozzle × Velocity]
α = Impact angle (rad)
R_a = Surface roughness (μm)
M = % of Moisture present in fly ash

Table 3.6 Comparison of the predicted exponents

	Present model	Literature
Impact velocity, Vp	0.88	1.5–3.2
Impact angle, α	−0.17	0.46–1.08
Fly-ash size, dp	0.3	0.33–3.9
Particle loading, Qp/V_F	0.5	–
Surface roughness, Ra	1.24	–
Moisture, M	−0.3	–
Titania, Ti	−0.4	–
Chromium, Cr	0.06	–

Ti = % of Titania present in fly ash
Cr = % of Chromium present in low alloy steel
a, b, c, d, e, f, g, h = Model parameters
Nozzle Diameter = 4.55 mm.

$$E = 7.08 \times 10^{-6} \frac{V_P^{0.8828} d_P^{0..3128} \left(\frac{Q_P}{V_F}\right)^{0.5075} (R_a)^{1.2431} (Cr)^{0.0693}}{(Sin\alpha)^{0.1787} M^{0.3265} (Ti)^{0.4105}} \quad (3.2)$$

Table 3.6 shows a comparison of the values of the various exponents derived from the present data with those reported in the literature [5]. It can be seen that there is considerable discrepancy between the two sets. Given that repeat experiments over a range of parameters have been carried out in the present study, this would mean that Indian coals exhibit a different behavior. A more wide-ranging study, focusing primarily on this issue, is therefore recommended.

The present results throw some light on the effect of the ash properties on the erosion rate. From the above model surface roughness, impact velocity, fly-ash particle size, % of titania in fly-ash and fly-ash loading are the most sensitive parameters to determine room temperature fly-ash erosion. Also, the erosion equation can be simplified as

$$E\, \alpha\, d_P^{0..3} \left(\frac{Q_P}{V_F}\right)^{0.5} \quad (3.3)$$

From Eq. 3.3, it appears that particle size (d_P) and the fly-ash particle loading (Q_P/V_F) are key factors in determining the rate of fly-ash impact metal erosion. This can be controlled by using ultrasonic method of coal-wash. Therefore, pre-combustion measures of coal beneficiation that result in size reduction and ash removal will lead to considerable mitigation of the fly-ash erosion problem. This forms part of the motivation for the present work on ultrasonic coal-wash method for this dual purpose.

3.4 Summary

The results of this work are summarized as follows:

- The rate of erosion increases with increasing impact velocity of fly-ash on metal.
- The erosion rate decreases monotonically with an increase in the impingement angle in the range of $15° < \Theta < 45°$, above 15° in the tested range.
- The erosion rate increases with an increase in particle size up to about 112.5 μm, and levels off after that size.
- The erosion rate increases with increasing concentration of fly-ash.
- Increasing surface roughness severely exacerbates erosive wear and mass loss.
- Hardness of the low-alloy steel (in the range tested) has only secondary influence on erosion, a trend consistent with previous observations in literature [6, 7]
- The % of titania present in the fly ash sample shows inverse effect on erosion rate. The effect of silica content could not be quantified as all coals tested were very similar in this critical parameter.

References

1. Alam HG, Mogaddam AG, Omidkhah MR (2009) The influence of process parameters on desulfurization of Mezino coal by HNO_3/HCl leaching. Fuel Process Technol 90:1–7
2. Jennings WH, Head WJ, Mannings CR Jr (1976) A mechanistic model for the prediction of ductile erosion. Wear 40:93
3. Das SK, Godiwalla KM, Mehrotra SP, Sastry KKM, Dey PK (2006) Analytical model for erosion behavior of impacted fly-ash particles on coal-fired boiler components. Sadhana 31(Part 5):583–595
4. Wang YF, Yang ZG (2008) Finite element model of erosive wear on ductile and brittle materials. Wear 265:871–878
5. Meng HC, Ludema KC (1995) Wear models and predictive equations: their form and content. Wear 181:443–457
6. Foley T, Levy A (1983) The erosion of heat-treated steels. Wear 91:45–64
7. Sundararajan G, Shewmon PG (1983) A new model for the erosion of metals at normal incidence. Wear 84:237–258

Chapter 4
Experimental Studies on Ultrasonic Coal Beneficiation

4.1 Overview

While ultrasonic coal-wash is not entirely new in many countries, it has not yet been practised in India, though it would appear that the relatively high ash and sulfur content of Indian coals would render them particularly suitable for such a washing procedure. In this study, state-of-the-art ultrasonic equipment, spanning the frequency range from, highly-cavitational (<100 kHz), to intermediate (100–200 kHz) to mostly acoustic-streaming dominated (>470 kHz), is used. The present study encompasses ultrasonic aqueous-based and reagent-based coal beneficiation, in addition to the wide frequency range employed. The uniqueness of the work results from these aspects.

4.2 Materials, Methods and Equipment Used

4.2.1 Materials

Coal is a combustible black or brownish-black sedimentary rock normally occurring in rock strata in layers or veins called coal beds or coal seams. The harder forms, such as anthracite, can be regarded as metamorphic rock because of later exposure to elevated temperature and pressure. Coal is composed primarily of carbon along with variable quantities of other elements such as S, H, O and N. As geological processes apply pressure to dead biotic material over time, under suitable conditions it is transformed successively into Peat and Lignite. The latter, also referred to as brown coal, is the lowest rank of coal and is used almost exclusively as fuel for electric power generation. Sub-bituminous coal, whose properties range from those of lignite to those of bituminous coal, which is used primarily as fuel for steam-electric power generation. Bituminous coal, dense sedimentary rock, black but sometimes dark

B. Ambedkar, *Ultrasonic Coal-Wash for De-Ashing and De-Sulfurization*,
Springer Theses, DOI: 10.1007/978-3-642-25017-0_4,

Table 4.1 Proximate analysis and total and forms of sulfur of as-received coal (wt%)

	Moisture	Volatile matter	Fixed carbon	Ash	Total sulfur	Pyritic + organic S	Sulphate sulfur
Lignite I	17.86	26.77	36.47	18.9	5.03	2.58	2.45
Lignite II	19.46	31.74	24.63	24.17	5.62	NM	NM
Belphar rejects	5.58	19.89	25.19	46.04	Not Measured (NM)		
Belphar washed	6.37	31.48	28.04	33.24			
Belphar ROM	6.83	34.63	22.79	34.29			
Dipka rejects	3.85	21.06	20.67	36.78			
Dipka washed	4.69	38.30	23.35	31.50			
Dipka ROM	6.28	19.46	33.02	38.18			
Talcher rejects	3.44	22.07	21.78	52.17			
Talcher washed	7.43	32.98	42.40	17.19			
Talcher ROM	2.6	23.75	31.01	40.64			

brown, often with well-defined bands of bright and dull material is used primarily as fuel in steam-electric power generation, with substantial quantities also used for heat and power applications in manufacturing and to make coke. Anthracite, the highest rank coal is a hard, glossy, black coal used primarily for residential and commercial space heating. Graphite, technically the highest rank, but difficult to ignite, is not so commonly used as fuel.

The coals that are used in this study and their properties are shown in Table 4.1. The proximate and sulfur (IS 1350) analysis of the coal samples were performed according to ASTM Standard procedures given Sect. 4.2.2. The first two coals are high-sulfur lignite. The sulfur content varies from 5 to 5.6 wt%. Moisture content of lignite coal is about 18%. The remaining nine types of coal are sub-bituminous coals. These coals are received from three different mines (Belphar, Dipka and Talcher mine), and each mine supplies three types of pre-processed coals namely, reject, conventionally washed and Run of Mine (ROM). Reject coals have too high an ash content. The ash content varies from 17 to 52%, fixed carbon 20–42%, volatile matter 19–38% and moisture varies from 2 to 7.5%. These nine types of coals are used for ultrasonic reagent-based de-ashing.

4.2.2 Methods

4.2.2.1 Determination of Ash Content (IS 1350)

Outline of the Method

The sample is heated in air to 500 °C in 30 min, from 500 to 815 °C for a further 30–60 min and is maintained at this temperature until constant in mass.

Apparatus

Muffle furnace was used for this analysis. The furnace is capable of being raised to a temperature of 850 ± 10 °C.

Procedure

The air-dried material was thoroughly mixed and ground to pass through a 212 μ IS sieve. The clean dry empty dish was weighed. Into this dish, 0.5 g of coal sample were taken and weighed accurately. Then, the material was distributed uniformly so that the spread does not exceed 0.15 g per cm^2. The uncovered dish was kept inside the muffle furnace at room temperature. The temperature was then raised to 500 °C in 30 min and to 815 ± 10 °C in a further 60 min, and was maintained at this temperature until the change in mass of the ash was less than 0.001 g. The dish was then removed from the furnace, covered with its lid, allowed to cool and weighed. The ash was brushed out and the empty dish reweighed.

Calculation

Ash, percent by mass = 100 × [(M3 − M4)/(M2 − M1)]

where

M1 = mass in g of dish,
M2 = mass in g of dish and sample,
M3 = mass in g of dish and ash, and
M4 = mass in g of dish after brushing out the ash and on reweighing.

4.2.2.2 Determination of Total Sulfur Content (IS 1350)

Outline of the Method

The sample of coal is heated in intimate contact with Eschka's mixture (mixture of two parts of MgO and one part of Na_2CO_3) in an oxidizing atmosphere to remove combustible matter and to convert the sulfur to sulphate. This is then extracted and determined by gravimetric method by precipitation as barium sulphate with barium chloride.

Procedure

Preparation of solution

The bottom of the 50 ml crucible is uniformly covered with 0.5 g of Eschka's mixture. The appropriate quantity of the material, crushed to pass 212 μ IS sieve, is weighed accurately and mixed intimately with 2.5 g of Eschka's mixture in an evaporating basin or other suitable vessel, and is brushed into the crucible. The content is leveled by tapping gently on the bench and is covered uniformly with one gram of Eschka mixture. The charged crucible is placed into the cold muffle furnace. The temperature is raised to 800 ± 25 °C in about one hour and then heated for a further 90 min. The plate is withdrawn with its crucible and cooled. The ignited mixture is transferred as completely as possible from the crucible to a beaker containing 25–30 ml of water. If unburnt particles are observed visually, the determination is rejected. The crucible is thoroughly washed out with about 50 ml of hot distilled water and the washings are added to the contents of the beaker.

Extraction by acid

A cover-glass is placed on the beaker, and then sufficient amount of concentrated hydrochloric acid (analytical grade) (17 ml will normally be required) is added to dissolve the solid matter, with the contents of the beaker being kept at warm condition to effect solution. Then, the solution is boiled for 5 min to expel carbon dioxide and filtered through a filter pad or medium-textured double acid-washed filter paper, the filtrate being collected in a 400 ml conical beaker. The residue on the filter pad or filter paper is washed with four 20 ml portions of hot distilled water. To the combined filtrate and washings, 2–3 drops of methyl red indicator are added; then, ammonium hydroxide solution is added cautiously until the color of the indicator changes and a trace of precipitate is formed. Sufficient amount of concentrated hydrochloric acid is added to re-dissolve the precipitate; subsequently, 1 ml is added in excess.

Precipitation of barium sulphate

The volume of the solution is made-up to approximately 300 ml with water. The beaker is covered and heated until the solution boils; then, the heating is slightly reduced until ebullition ceases. 10 ml of barium chloride solution is added within approximately 20 s from a pipette held such that the barium chloride falls into the middle of the hot solution which is being agitated. The solution is kept just below boiling point, for 30 min.

Filtration

The barium sulphate is recovered by gravity, through an ash-less fine-textured double acid-washed filter paper in a long-stemmed 60° funnel (Whatman No. 42 filter paper is suitable for this purpose). Then, it is dried and weighed.

Calculation

Sulfur, percent by weight $= [13.74 \times (A - B + 0.008)/W]$

Where,

A = weight in grams of barium sulphate found in the determination,
B = weight in grams of barium sulphate found in the blank determination, and
W = weight in grams of the material taken for the test.

4.2.3 Equipment Used

4.2.3.1 Schematic of Ultrasonic System

The ultrasonic system (Fig. 4.1a) has three components, namely, an ultrasonic generator, an ultrasonic transducer and a tank with liquid. The ultrasonic tank is a bright-annealed stainless steel tank that has piezo-electric transducers mounted at the bottom. The ultrasonic generator transforms the line voltage to a frequency corresponding to the operating frequency of the transducer. The transducer transforms these electric oscillations to mechanical sound waves.

Figure 4.1b shows schematic view of probe-type ultrasonic system. In the probe-type, horn tip vibrates and generates ultrasonic waves in liquid present in a vessel. During propagation of wave, cavitations in liquids and bulk fluid motion

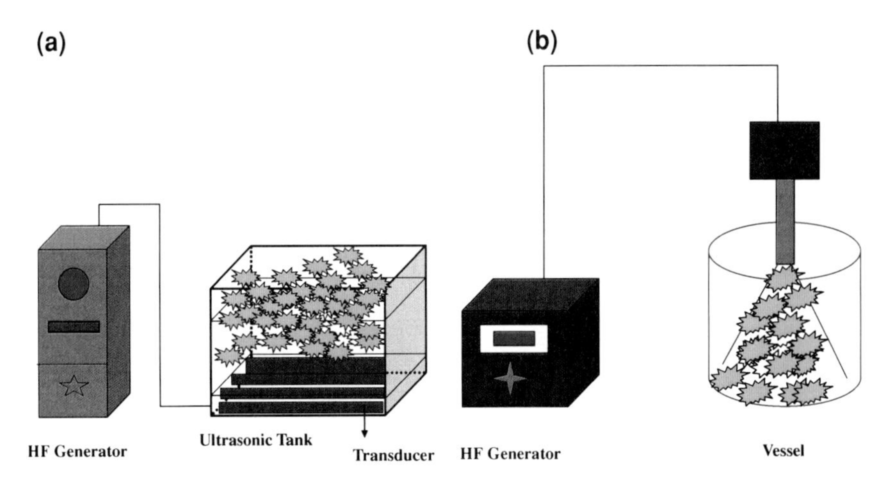

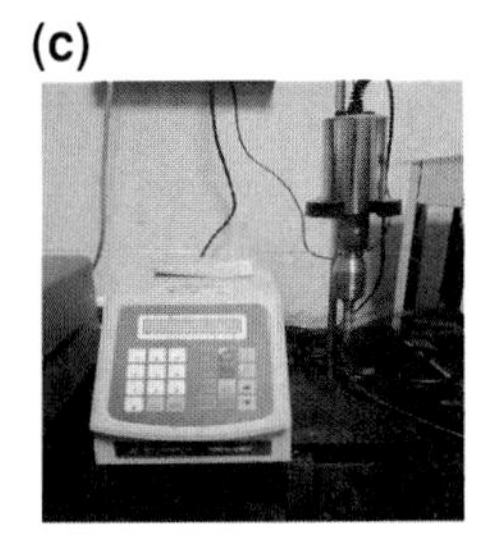

Fig. 4.1 **a** Schematic of tank-type ultrasonics. **b** Schematic of probe-type ultrasonics. **c** Photographic view of probe-type ultrasonics

due to streaming are formed. Photographic view of probe-type ultrasonics is shown in Fig. 4.1c.

Figure 4.2 shows fluid behavior at various ultrasonic frequencies. At 25 kHz, the liquid medium appears stagnant, but is experiencing very high intense cavitation in the tank. The stagnancy of the liquid is evidence that the 25 kHz frequency is only cavitation-dominant. The liquid behavior in the dual-frequency (58/192) tank is somewhat different, resembling a slithering movement. In the 430 kHz tank, acoustic streaming is the dominant phenomenon. This is confirmed by "fountain" effect seen in the center of the tank.

Energy dissipated as heat in an ultrasonic irradiation is calculated using calorimetric study [1, 2] and is shown in Table 4.2. The energy dissipation is minimum in the case of tank type ultrasonic system.

4.2.3.2 Laser Particle Counter (LPC)

Spectrex Laser Particle Counter (Fig. 4.3) is used to measure the particle size in the range of 1–100 μ. Utilizing the principle of "near angle light scatter", a revolving laser beam passes through the walls of a glass container of a flow-

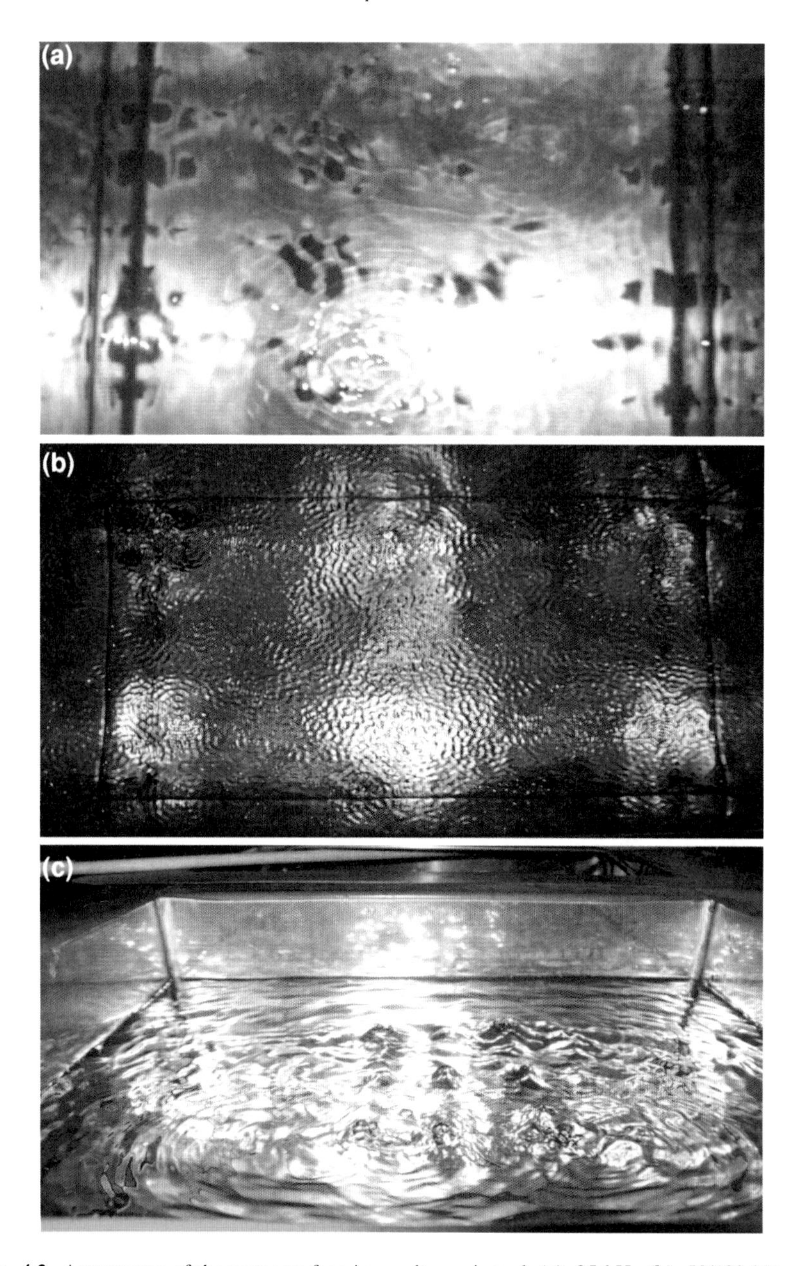

Fig. 4.2 Appearance of the water surface in an ultrasonic tank (**a**), 25 kHz (**b**), 58/192 kHz and (**c**), 430 kHz. The power input is 500 W in all cases

through cell. When it is directed through a central "sensitive zone", the PC-2200^R not only counts the particles in suspension, but tabulates their size as well. The analog signals generated by the light pulses are routed to a computer and digitized.

Table 4.2 Power dissipated as heat by calorimetric study

Frequency (kHz)	Power input (W)	Slope (°C/s)	Power dissipated as heat (W)
25	500	0.0056	11.72
58/192	500	0.0052	10.88
430	500	0.0086	18.00
20kHz_Probe	500	0.0261	54.63

4.3 Experimental Procedure

4.3.1 Aqueous-Based De-Ashing Experiments

A 20 kHz probe and 25, dual (58/192) and 430 kHz ultrasonic tanks have been used for this investigation. The coal (Lignite I, ref Table 4.1) to be studied was received from Giral mine, Rajasthan, India. The coal sample was air dried and ground using ball mill. Then, it was sieved using Indian Standard sieves to required particle size. The proximate and sulfur analysis of the coal samples were performed according to ASTM standard procedures and are given in Table 4.1.

20 g of 212 μ sieve pass-through lignite coal I and 500 ml of water were taken in a 1 L borosil beaker. The mixture was subjected to sonication (20 kHz Probe, 25, 58/192 and 430 kHz ultrasonic tank) for about 15 min, with 500 W input power of ultrasound. Then, the sonicated sample was separated into three levels by decantation method. In order to separate the detached ash from coal, a settling column has been fabricated. It is of 1 m height, and 15 cm in diameter, having 3 outlets as shown in Fig. 4.4a and b and the column is filled with water. The sonicated coal slurry may be passed through a settling column, and 2 min of settling time was provided for the particles to settle. Coal, being lighter than ash material, will have low settling velocity and will settle slower, facilitating separation. Then the sonicated sample was separated into three levels by decantation. The top level sample is collected first, next middle level coal sample finally the bottom of the coal sample. The three levels of coal sample were filtered, washed with water and dried at 100 °C for about 5 h. The dried coal samples were subjected to ash analysis for ash content, and to SEM analysis to examine surface morphology of virgin and sonicated coal sample.

4.3.2 Aqueous-Based De-Sulfurization Experiments

These experiments were conducted for 20 g of 212 μm sieve pass-through high sulfur lignite coal I in 500 ml of water using probe-type and tank-type sonicator. Probe-type sonicator of 20 kHz frequency or a tank-type sonicator of 25, 58/192

Fig. 4.3 Photographic view of laser particle counter, model no. PC-2200

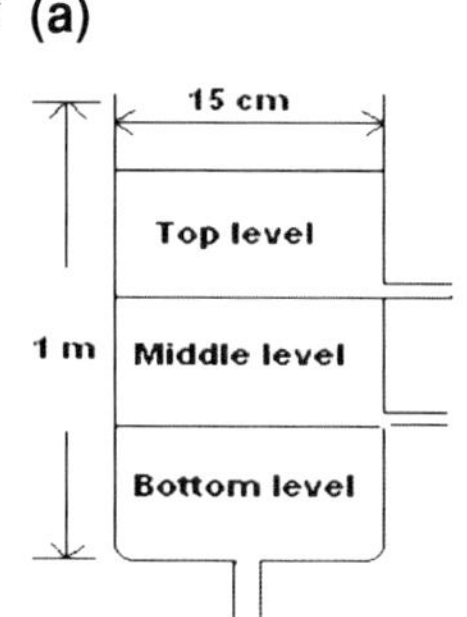

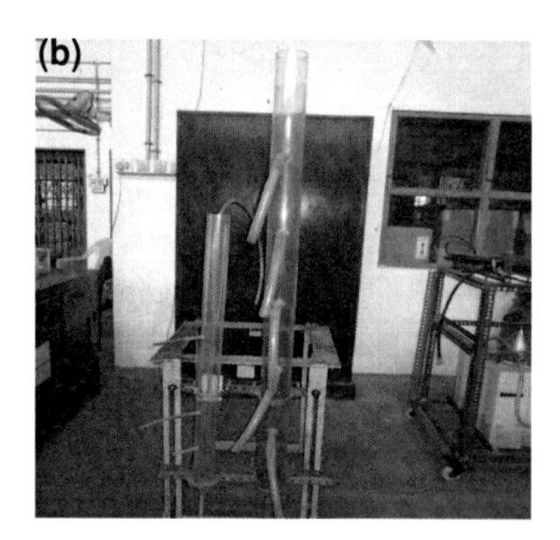

Fig. 4.4 **a** Schematic of decanting column. **b** Photographic view of decanting column

(dual) and 430 kHz frequencies was used. Experiments were conducted for about 60 min and 500 W power. After sonication, the treated sample mixture was filtered, washed with water and dried for about 5 h in an oven at 100 °C. The treated coal sample was analyzed for sulfur content according to the procedure given in Sect. 4.2.2.2.

4.3.3 Solvent-Based De-Ashing Experiments

Coals from three different mines (Belphar, Dipka, Talcher), and in three different as-received conditions—run-of-the-mine (ROM), washed (conventionally washed) and rejects—were used in this study. Proximate analyses of the coals are reported in Table 4.1. The coal samples used in this study were crushed using a Hazemag crusher. Then, the samples were split into three different size ranges: (− 1 + 0.09), (− 2 + 1) and (− 4 + 2) mm. Samples were then dried. Taguchi's parameter design provides a simple and systematic approach for optimization of design for quality, performance and cost. Table 4.3 shows the factors and levels for L_{27} DOE. To investigate reagent-based de-ashing with five controllable three-

Table 4.3 Taguchi L_{27} DOE: parameters and levels for De-Ashing experiments

Parameter	Level 1	Level 2	Level 3
Ultrasonic frequency (kHz)	25	132	470
Coal/methanol ratio (by volume)	1:6	1:4	1:2
Initial temperature (°C)	25	35	45
Coal condition	ROM	Washed	Reject
Initial coal size (mm)	(− 1 + 0.09)	(− 2 + 1)	(− 4 + 2)

level factors, an L_{27} array including 27 tests was chosen and is shown in Table 4.4. L_{27} DOE is repeated for three different mines (Belphar, Dipka and Talcher) of coals.

De-ashing and particle size-reduction experiments were carried out as follows. Each run was conducted for about 15 min and at 500 W power. 10 g of coal of known size range were taken in a 150 ml beaker, and solvent (methanol) was poured into the beaker, as per the trial condition specified in DOE. The mixture was subjected to sonication. Then, the sonicated sample mixture was filtered. The filtered coal was washed with water and dried for 5 h in an oven at 100 °C. The dried coal sample was subjected to sieve analysis; the same set of mesh was used for post sieving analysis. The weight % of coal particles passing through the lowest mesh was considered to calculate % size reduction. The treated coal was analyzed for ash content according to the procedure given in Sect. 4.2.2.1.

4.3.4 Reagent-Based De-Sulfurization Experiments

212 μm sieve passing-through high-sulfur lignite coal I was taken for desulfurization experiments. The proximate and sulfur analysis of the coal samples was performed using IS 1350 procedure and shown in Table 4.1. Desulfurization experiments were carried out by conventional soaking, agitation and ultrasonic methods. 2N HNO_3 and 3-volume percentage H_2O_2 were used as a reagent. The reagents used were of analytical grade. The experiments were conducted using a beaker and non-adiabatic condition. The mixture was subjected to soaking/stirring/sonication (20 kHz Probe). One end of the thermocouple was immersed into the coal-reagent mixture; the other end was connected to a data-logger to monitor the reaction temperature at regular intervals during sonication. The prime reason for measuring mixture temperature is to track the rate at which the reaction proceeds during soaking, stirring and ultrasound irradiation. Then, the treated sample mixture was filtered, washed with water and dried for about 5 h in an oven at 100 °C. The treated coal sample was analyzed for sulfur content according to the procedure given in Sect. 4.2.2.2

Table 4.4 L_{27} Orthogonal array for De-Ashing experiments

Trial no.	Frequency, KHz	Initial coal size, mm	Coal-solvent ratio, vol%	Initial temp, °C	Coal type
1	25	(−212 + 0)	1:6	25	Reject
2	25	(−212 + 0)	1:6	25	Washed
3	25	(−212 + 0)	1:6	25	ROM
4	25	(−2 + 1)	1:4	35	Reject
5	25	(−2 + 1)	1:4	35	Washed
6	25	(−2 + 1)	1:4	35	ROM
7	25	(−4 + 2)	1:2	45	Reject
8	25	(−4 + 2)	1:2	45	Washed
9	25	(−4 + 2)	1:2	45	ROM
10	132	(−212 + 0)	1:4	45	Washed
11	132	(−212 + 0)	1:4	45	ROM
12	132	(−212 + 0)	1:4	45	Reject
13	132	(−2 + 1)	1:2	25	Washed
14	132	(−2 + 1)	1:2	25	ROM
15	132	(−2 + 1)	1:2	25	Reject
16	132	(−4 + 2)	1:6	35	ROM
17	132	(−4 + 2)	1:6	35	Reject
18	132	(−4 + 2)	1:6	35	Washed
19	470	(−212 + 0)	1:2	35	ROM
20	470	(−212 + 0)	1:2	35	Reject
21	470	(−212 + 0)	1:2	35	Washed
22	470	(−2 + 1)	1:6	45	ROM
23	470	(−2 + 1)	1:6	45	Reject
24	470	(−2 + 1)	1:6	45	Washed
25	470	(−4 + 2)	1:4	25	ROM
26	470	(−4 + 2)	1:4	25	Reject
27	470	(−4 + 2)	1:4	25	Washed

1. **Soaking**

Soaking experiments were conducted using 2N HNO_3 and 3-volume percentage H_2O_2. 10 g of 212 μm pass-through high-sulfur coal I sample were soaked in 100 ml of 2N HNO_3 and 3-volume percentage of H_2O_2 for 5 h.

2. **Stirring**

10 g of 212 μm pass-through high-sulfur lignite coal I and 100 ml of 2N HNO_3 and 3-volume percentage of H_2O_2 were taken for the conventional stirring process. Stirring was conducted for about 1 h at the speed of 1,000 rpm.

3. **Sonication**

In order to see the effect of ultrasonics on reagent-based desulfurization, 20 kHz probe is used at 500 W input power. 10 g of 212 μm sieve pass-through high-sulfur lignite coal I and 100 ml of reagents were taken in a 250 ml beaker. The reaction mixture was sonicated for intervals of 10, 20 and 30 min.

4.4 Results for Aqueous-Based Coal Beneficiation

4.4.1 Motivation

Aqueous-based coal beneficiation using ultrasound is attributable to two basic mechanisms: 1. Coal particle breakage, and 2. Leaching. Ultrasound-assisted particle breakage is entirely different from conventional particle breakage. The interaction mechanism between suspended coal particle and ultrasound, and mechanism of coal particle breakage and ash removal, are therefore investigated here.

4.4.2 De-Ashing

4.4.2.1 Ultrasound Assisted Coal Particle Breakage

In low-frequency ultrasound (<100 kHz), cavitation phenomena are predominant, and particle breakage is extensive. In high-frequency ultrasound (>100 kHz), streaming phenomena are dominant, and leaching effect prevails. Cavitation is due to implosion of bubbles in the acoustic field, and to the resulting transmission of a shock wave. Millions of bubbles will form, grow and collapse within a nanosecond. The collective effect due to bubble implosion will be enormous (1,000's of K temperature and 100's of atm pressure).

The effect of cavitation is several hundred times greater in heterogeneous than in homogeneous systems [3]. Unlike cavitation bubble collapse in homogenous systems (liquid–liquid interface), collapse of a cavitation bubble in heterogeneous systems (e.g., liquid–solid) on or near to a surface is non-symmetrical in nature since the surface provides resistance to liquid flow. The result is an in-rush of liquid predominantly from the opposite side of the bubble, resulting in a powerful liquid jet being formed and targeted at the surface. It is also important to note that the rapid collapse of the cavitation bubbles generates significant shear forces in the bulk liquid immediately surrounding the bubbles and, as a result, produces a strong stirring mechanical effect. These effects can significantly increase mass transfer to the surface [4]. Cavitation is also important in case of heterogeneous systems in that most of the cavitation bubbles are generated close to the surface of the substrate, thus providing an important additional benefit of the "opening up" of the surface of solid substrates as a result of mechanical impacts produced by powerful "jets" of collapsing cavitational bubbles.

This will cause particle breakage in different ways: pitting of coal particle surface to produce fines, and forming cracks on the surface, which are widened and deepened due to prolonged exposure, finally causing breakage. Coal particle breakage mechanism is illustrated here using lignite I coal sample sonicated for 5 min by 25 kHz frequency of ultrasound, 500 W input powers. SEM images of virgin and sonicated coal samples are shown in Fig. 4.5.

Four stages for ultrasound assisted coal particle breakage can be observed here. These are:

1. Pitting of the coal surfaces (Fig. 4.5b)
2. Formation of cracks on the coal surface (Fig. 4.5c)
3. Widening and deepening of coal surface cracks (Fig. 4.5d) and
4. Breakage of coal particles (Fig. 4.5e).

4.4.2.2 Size Distribution Analysis of Virgin and Sonicated Sample

Size distribution analysis of virgin and sonicated coal sample at various ultrasonic frequencies is shown in Fig. 4.6. Figure 4.6a shows size distribution analysis of virgin, 2 min and 5 min sonicated coal samples by 20 kHz probe type ultrasonics. From the data, it can be observed that a 2 min sonication yields an entirely new spectrum of particle sizes, while a 5 min sonication yields another new size spectrum, overlapping the previous one to some extent. This suggests that, even with probe-type ultrasonics directly immersed in the coal slurry, particles which are in the vicinity of the probe experience higher intensity of ultrasound compared to more remote particles. But in the case of 25 kHz tank (Fig. 4.6b), system provides uniform energy transfer to the coal slurry, yielding an entirely new size spectrum with respect to sonication time. Dual (58/192 kHz) system (Fig. 4.6c) causes less particle breakage than the 25 kHz system due combination of low- and high-frequency ultrasound providing a combined effect of cavitation and streaming. In the case of 430 kHz tank system (Fig. 4.6d), coal particle breakage is negligible due to pure streaming phenomena being dominant in high frequency ultrasound.

Figures 4.7a, b show effect of ultrasonic frequency on coal particle size reduction at various time intervals. Figure 4.7a shows size distribution analysis of virgin and 2 min sonicated coal sample by 25 kHz, dual (58/192 kHz) and 430 kHz. From the graph, it may be observed that 2 min of sonication at 25 kHz frequency yields an entirely new spectrum of particle sizes. Dual (58/192 kHz) system causes less particle breakage than the 25 kHz system, but there is an additional streaming effect due to 192 kHz frequency. In 430 kHz system, size reduction is negligible. Figure 4.7b shows size distribution analysis of virgin and 5 min sonicated coal samples by 25 kHz, dual (58/192 kHz) and 430 kHz. 5 min of sonication yields another new size spectrum with respect to ultrasonic frequency. This indicates that ultrasonic size reduction is a time-dependant phenomenon, and progresses rapidly towards smaller sizes. It has been observed that lower frequency leads to highest size reduction within a short duration of sonication. Higher frequency causes less particle breakage due to streaming phenomena being dominant in the system.

Figure 4.8 shows effect of ultrasonic frequency on the mean coal particle size at various times of sonication. It has been observed that lower frequency ultrasonics (20 kHz probe and 20 kHz tank) induce drastic change in the mean size during

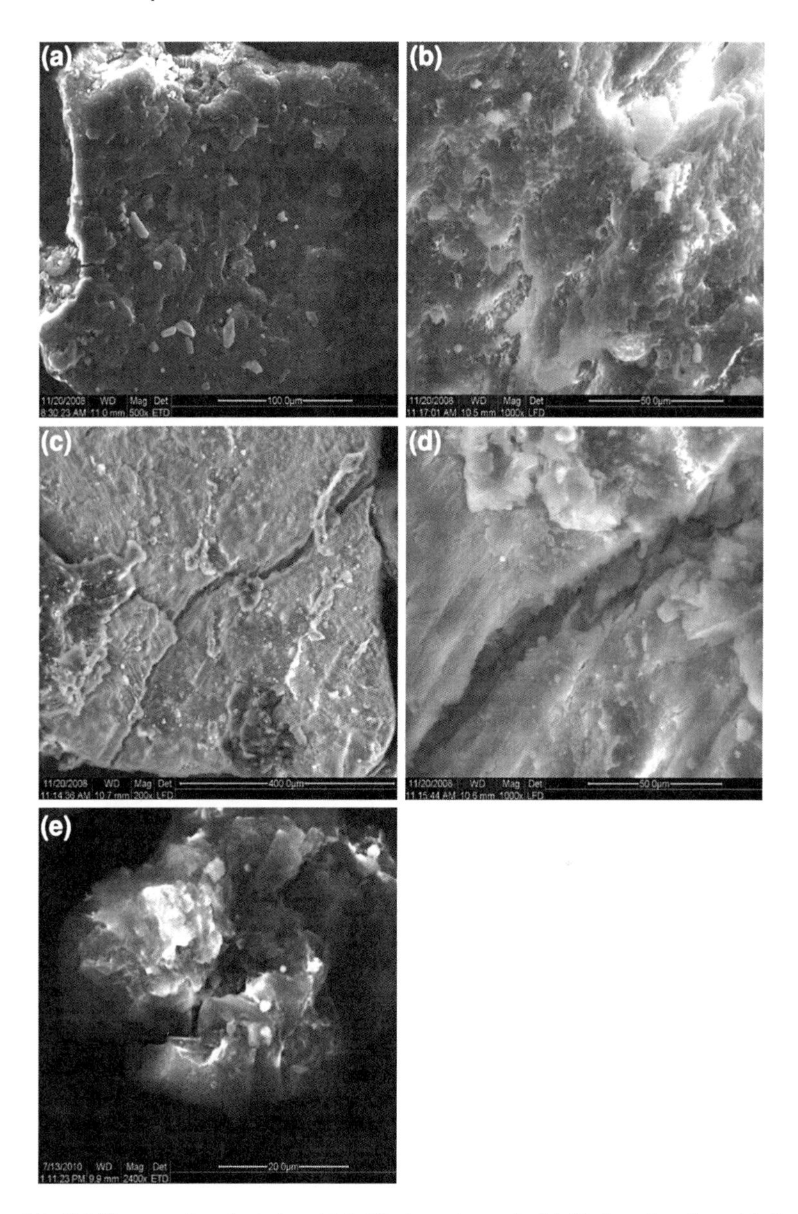

Fig. 4.5 SEM images of sonicated coal (**a**), Virgin coal sample (**b**), Pitting of surfaces (**c**), Crack formation (**d**), Widened cracks and (**e**), Breakage of particle

short period of sonication, leading to increasing total specific surface area, as shown in Fig. 4.9. The total specific area is calculated by $A_{sp} = \frac{6}{d_{pf}\phi_s\rho_p}$; Sphericity (ϕ_s) of coal particle is taken as 0.75. Density (ρ_p) is density of coal particle is taken as 753 kg/m^3. This investigation suggests that choice of ultrasonic

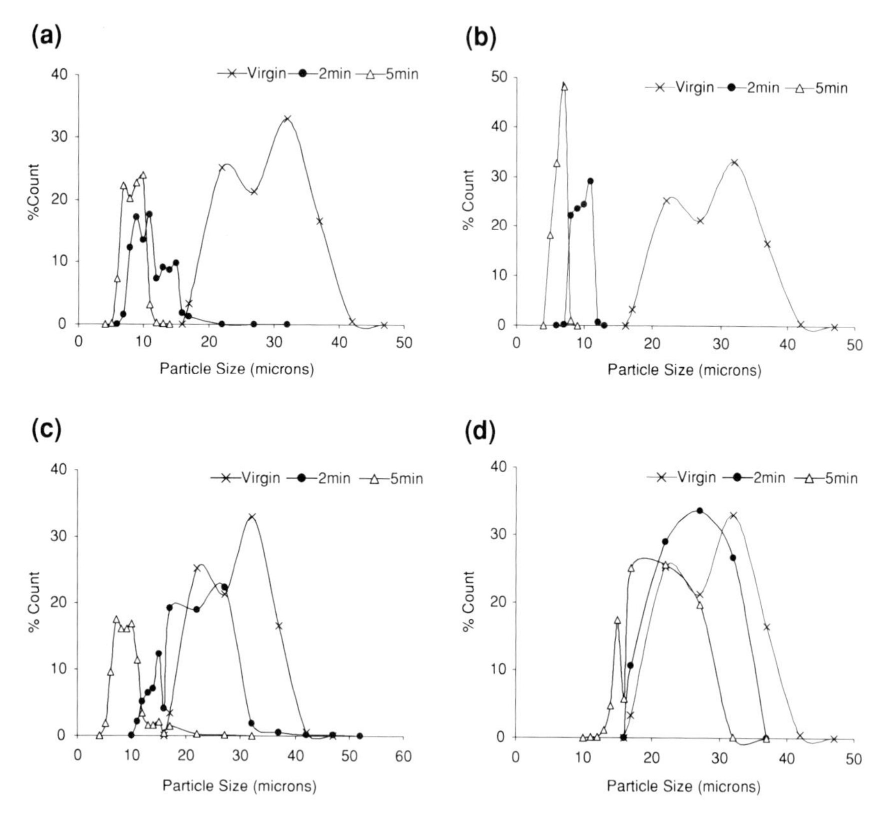

Fig. 4.6 Size distribution analysis of virgin and sonicated coal sample at various ultrasonic frequencies. **a** 20 kHz Probe. **b** 25 kHz Tank. **c** 58/192 kHz Tank. **d** 430 kHz Tank

frequency may be based upon desirable breakage characteristics. An optimum setting may be low-frequency for particle breakage, high frequency for pure leaching process and combination of low and high frequency for combined effect.

4.4.2.3 Ash Analysis of Virgin and Treated Coal Samples

Table 4.5 shows ash analysis of virgin and treated coal samples. The sonicated sample was separated into three levels by decantation. Top level was expected to be rich in lighter impurities, middle level was expected to be mostly clean coal; and bottom level was expected to be ash-rich coal (heavier impurities).

To confirm these trends, a semi-quantitative ash analysis has been done on three levels of coal samples. Semi-quantitative analysis in the sense that, the total mass conservation studies in the each level is excluded instead, part of coal samples picked up from the three levels for ash analysis. From Table 4.5, it can be confirmed that the ash content of the top and middle level is lower than that of the bottom level and

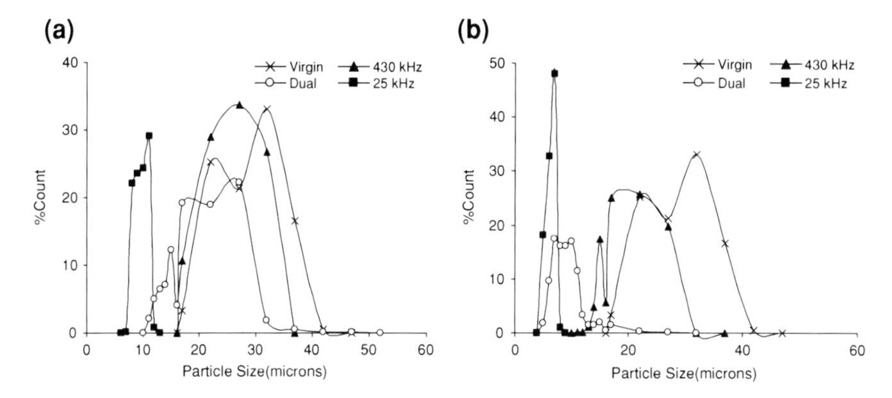

Fig. 4.7 Size distribution analysis of coal at various frequency of ultrasound. **a** 2 min sonication. **b** 5 min sonication

Fig. 4.8 Effect of ultrasonic frequency on coal particle size at various time intervals

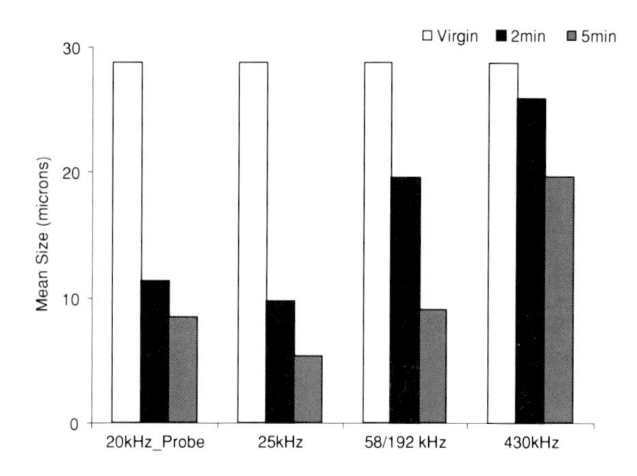

Fig. 4.9 Effect of sonication time on total specific surface area for different frequency of ultrasound

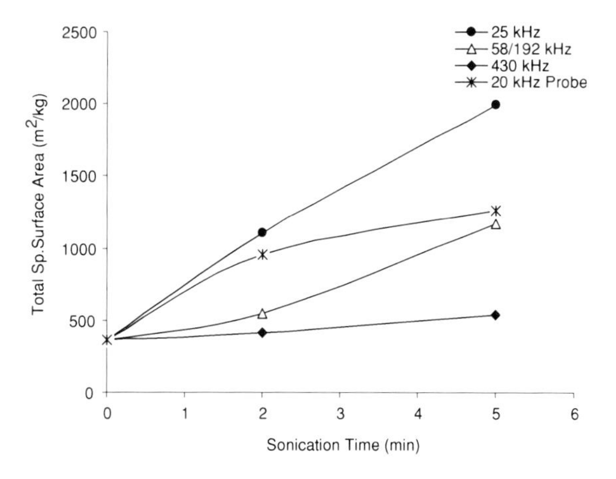

Table 4.5 Ash analysis of virgin and treated coal samples

	Virgin, %	Top, %	Middle, %	Bottom, %
Ball milling	18.9	18.61	19.12	22.35
20 kHz Probe	18.9	6.73	14.66	21.82
25 kHz Tank	18.9	9.80	13.36	21.07
58/192 kHz Tank	18.9	8.13	15.34	20.61
430 kHz Tank	18.9	15.26	17.78	21.02

virgin coal. Sono-fragmentation of coal leads to the detachment of ash impurities from coal and separation by decantation process using density difference as a driving force. In these experiments, 35–40% of coal samples are recovered from top and middle level, remaining are settled in the bottom at a given settling time. The bottom level coal sample requires further sonication for de-ashing.

4.4.2.4 SEM Analysis of Decanted Coal Sample: Dual Frequency (58/192 kHz)

To have a better understanding regarding distribution of minerals, fracture morphology, and impurity removal, virgin and sonicated coal sample [using dual (58/192 kHz) frequency] were examined through scanning electron microscope (SEM). Figure 4.10 shows SEM images of different levels of coal sample. In virgin coal sample, the inorganic minerals are distributed all over the sample (Fig. 4.5a). The inorganic impurities adhere to the coal in a rigid as well as loose manner. These impurities are removed by ultrasonication. Three levels of decanted coal sample were examined through scanning electron microscope. The SEM image of top level coal sample is shown in Fig. 4.10a. The fines produced by the ultrasound are taken from the top level sample, and look amorphous in structure. Figure 4.10b shows the SEM image of the middle-level coal sample SEM image. In that image, a number of unfilled cavities are seen. It is apparent that the unfilled cavities are due to removal of ash material by ultrasonication. This has been confirmed by ash analysis as well as EDAX elemental analysis (Tables 4.5 and 4.6). SEM image of bottom-level coal sample is shown in Fig. 4.10c. There is still some mineral matter (luminous part) left on the coal surface; this has been confirmed again by ash as well as EDAX elemental analysis. Hence, bottom-level coal sample is verified to require further sonication for improved de-ashing.

4.4.2.5 EDAX Analysis of Decanted Coal Sample: Dual Frequency (58/192 kHz)

Elemental analysis of virgin and three levels of decanted coal sample are shown in Table 4.6. EDAX analysis (semi-quantitative) has been done on virgin and part of coal sample picked up from each levels of decanting column to determine the weight % of elements in it.

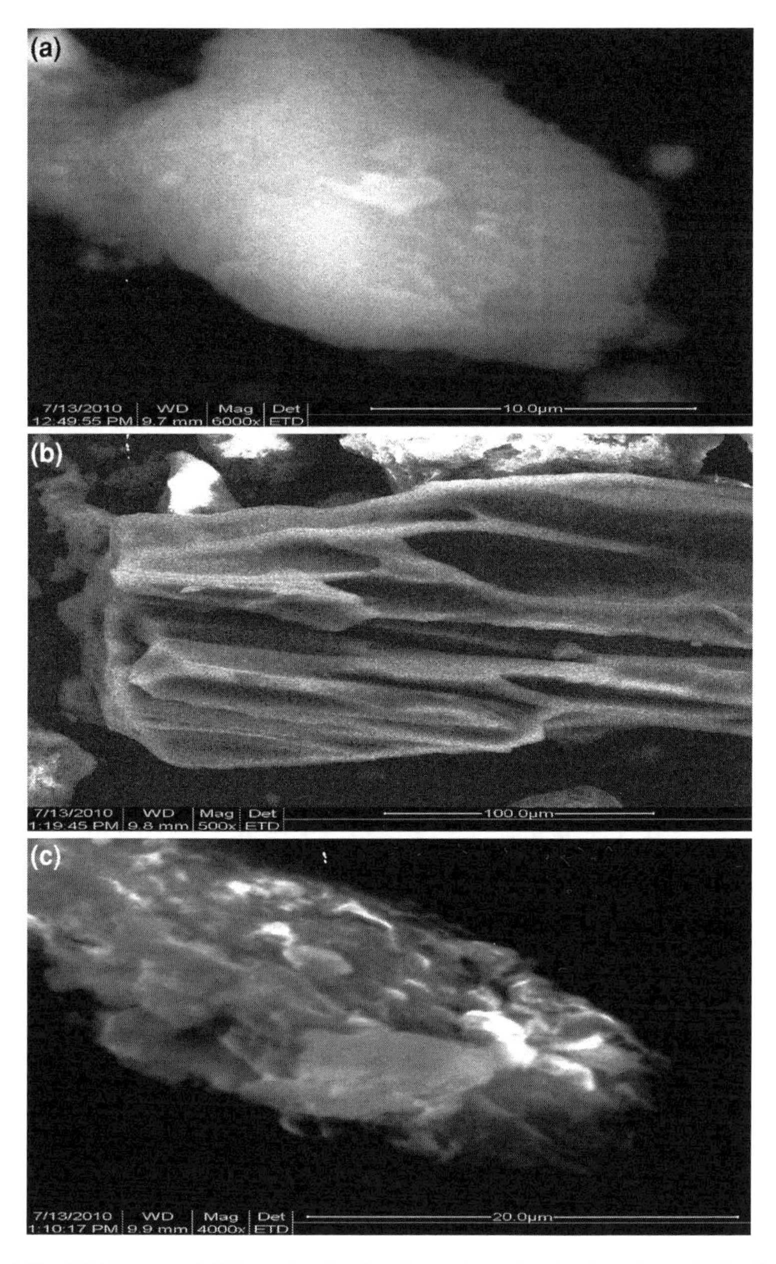

Fig. 4.10 SEM images of different levels of coal sample. **a** Top-level coal sample. **b** Middle-level coal sample. **c** Bottom-level coal sample

From Table 4.6, it is apparent that the weight % of carbon in middle level samples is higher than in other two levels and in the virgin coal sample. It is evident that, during sonication process, ash impurities were detached from the

Table 4.6 Elemental analysis of virgin and three levels of decanted coal sample

Elements	Virgin, %	Top, %	Middle, %	Bottom, %
C	58.41	75.22	87.26	31.62
O	27.68	19.07	10.90	33.69
Na	0.18	0.17	0.06	0.97
Mg	0.21	0.25	0.18	0.76
Al	2.84	0.91	0.13	10.68
Si	3.55	0.44	0.1	15.80
S	2.41	2.14	0.65	1.70
Cl	0.29	0.08	0.26	0
K	0.2	0	0	0.49
Ba	1.26	0.38	0.18	1.34
Fe	2.97	1.35	0.27	2.94

surface and interior part of the coal, and then separated by decantation process. The main mineral constituents are silica and alumina, which were observed in our analysis along with traces of iron and sulfur. From Table 4.6, the weight % of alumina, silica and iron are seen to be less in top and middle level than in virgin and bottom level coal sample. Size distribution by level has not been measured. One would expect that coal-rich, and therefore less dense, particles will be predominant in the top layer, and ash-rich, and therefore denser, particles in the bottom level. Further studies are required to ensure the mass conservation of each element in each level of the coal sample.

4.4.2.6 Summary

- Four stages for ultrasound assisted coal particle breakage have been proposed and substantiated with SEM images of sonicated coal. Pitting of coal surfaces is due to impingement of micro jets that are produced by cavitation bubble collapse. The crack formation, widening and deepening on the coal surface are due to subsequent collapse of cavitation bubbles.
- Ultrasonic size reduction is a time-dependant phenomenon, and progresses rapidly towards smaller sizes. Lower frequency leads to highest size reduction within short duration of sonication due to strong cavitational effect. Higher frequency causes less particle breakage due to streaming phenomena dominant in the system.
- Top-level decanted sample is mainly made up of lighter impurities, middle-level is rich in coal particle and bottom-level sample is rich in ash materials. This has been confirmed by ash analysis and EDAX elemental analysis.
- In virgin coal sample, the inorganic minerals (alumina, silica, iron and some traces) are distributed all over the sample, as can be seen in SEM images. Top-level sample shows an amorphous structure. Numerous unfilled cavities can be seen in middle-level coal sample due to removal of tightly bounded inorganic impurities (ash) by sonication. Bottom-level samples were rich in ash materials. Luminous feature due to ash (alumina) materials was observed.

4.4.3 De-Sulfurization

The main effects of ultrasound in liquid medium are acoustic cavitation and acoustic streaming. The process of formation, growth and implosion of bubbles is called cavitation. Bulk fluid motion due to sound energy absorption is known as acoustic streaming. In addition, coupling of an acoustic field to water produces OH radicals, H_2O_2, O_2, ozone and HO_2 that are strong oxidizing agents. Oxidation that occurs due to ultrasound is called Advanced Oxidation Process (AOP). It converts sulfur from coal to water-soluble sulphates (www.solvayinterox.com).

4.4.3.1 Effect of Ultrasonic Frequency on Total Sulfur Removal

The formation of H^+ and OH^- is attributed to the thermal dissociation of water vapor present in the cavities during the compression phase. Sonolysis of water also produces H_2O_2 and hydrogen gas, via hydroxyl radicals and hydrogen atoms [5]. The presence of oxygen improves sonochemical activities, but it is not essential for water sonolysis, and sonochemical oxidation and reduction can proceed in the presence of any gas. However, in the presence of oxygen acting as a scavenger of hydrogen atoms, the hydro-peroxyl radical is additionally formed and acts as an oxidizing agent. This radical causes a number of other reactions to occur, resulting in the formation of H_2O_2, O_2, O, and H_2 as products. Thermal dissociation of oxygen molecule may also occur, leading to the generation of additional hydroxyl radicals. In the absence of OH scavengers, the main product of the sonolysis of water is H_2O_2. H_2O_2 can also be produced in an "inert" atmosphere but only at the expense of OH radicals. These highly reacting species are responsible for coal desulfurization.

When a liquid is irradiated by ultrasound, cavitation will appear when the pressure amplitude of the applied ultrasound reaches a certain minimum. If it happens on the surface of the liquid, it consumes atmospheric air due to low pressure in an ultrasonic field. Atmospheric air consists of O_2, N_2, and some trace gases. The nitrogen tends to undergo advanced oxidation process and form NO, NO_2, and HNO_3. This has significant effect when the experiments are conducted in a large scale, the reason being that the surface available for absorbing N_2 from atmosphere is large. But, for experiments conducted in a laboratory scale, surface available for absorbing atmospheric N_2 is small; hence, HNO_3 formation and consequent enhancement of desulfurization of coal has insignificant effect.

Figure 4.11 shows effect of ultrasonic frequency on total sulfur removal. The sulfur removed using 25 kHz ultrasonic system is mainly because of acoustic cavitation, which produces high surface area of particle (Fig. 4.9) leading to intimate contact with the strong oxidizing agents (OH radicals, H_2O_2, ozone) produced in the ultrasonic fields, which convert the coal sulfur into water soluble sulphates (www.solvayinterox.com). Figures 4.7a, b show the size distribution of

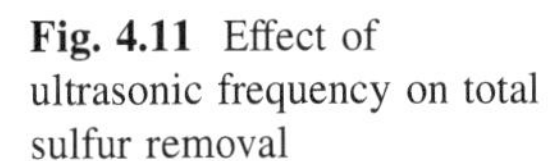

Fig. 4.11 Effect of ultrasonic frequency on total sulfur removal

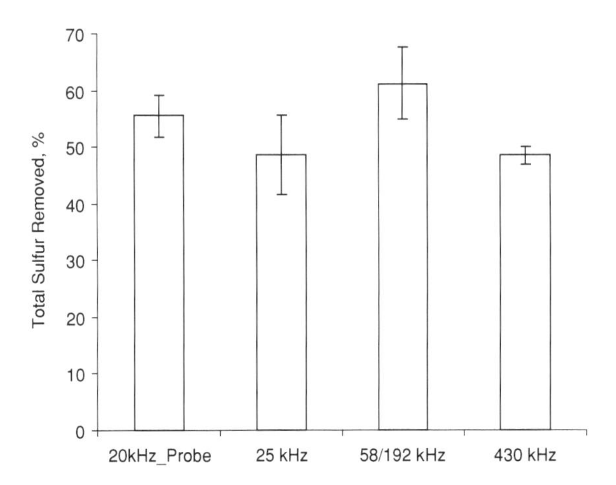

virgin and sonicated coal samples with two different ultrasonic frequencies −20 kHz probe, and 25 kHz ultrasonic tank. Breakage was more predominant in the low-frequency ultrasound field. It produces very fine particles within short period (2 and 5 min) of sonication, thereby increasing surface area of coal particle, causing it to have an intimate contact with the strong oxidizing agents produced by the ultrasound in an aqueous medium. The sphericity of the particle also increases with sonication time, due to the associated micro-polishing mechanism [6]. In the 20 kHz probe-type sonicator, cavitation is dominant, and the probe is in direct contact with the sample mixture. Therefore, particle breakage is more pronounced than in the other four cases. It leads to higher sulfur removal compared to the other three ultrasonic tank systems with same power input. But, the main disadvantage of the probe-type ultrasonic system is non-uniformity in energy transferred, which can be clearly understand from the effect of probe-type sonication on particle size distribution (Fig. 4.7a) suggesting that size spectrum of 5 min sonication is overlapping with the previous size spectrum (2 min sonication). The large amount of input power dissipated as heat is shown in Table 4.2. In the 430 kHz ultrasonic field, acoustic streaming phenomena are predominant. Leaching of impurities from coal is induced by streaming effect. Hence, the total sulfur removed in coal using 430 kHz ultrasonic system is mainly because of streaming effect. In dual system, the cavitation present due to 58 kHz leads to particle breakage, while the streaming associated with 192 kHz leaches out the contaminants. This dual mechanism effect results in highest removal of sulfur.

4.4.3.2 Effect of Ultrasonic Frequency on Removal of Various Forms of Sulfur

In coal, sulfur exists in three different forms: 1. sulphate sulfur, 2. pyritic sulfur, and 3. organic sulfur. Analysis has been performed on ultrasonically treated samples to

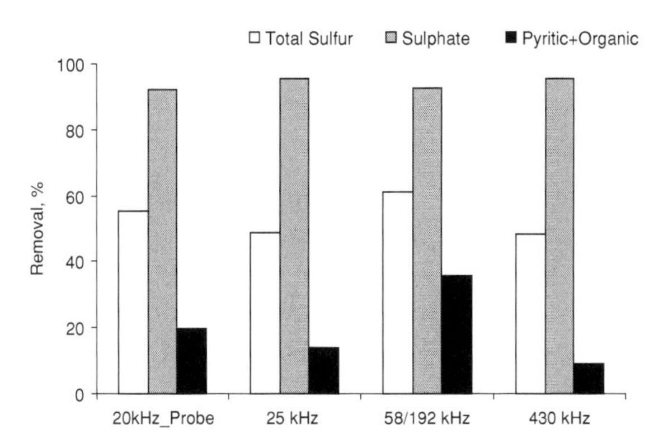

Fig. 4.12 Effect of frequency on different forms of sulfur removal

see how the forms of sulfur are distributed after ultrasonic treatment. From Fig. 4.12, it is apparent that almost all the cases show 90% and above sulphate sulfur removal. However, the pyritic and organic sulfur shows high resistance to ultrasonic treatment in water. 20 kHz probe systems remove almost 20% of pyritic + organic sulfur due to high energy transferred to the sample mixture. The 25 kHz ultrasonic tank system removes 14% of pyritic + organic sulfur, whereas the 430 kHz system removes only about 9% due to a single relatively-mild mechanism being in effect. Dual system removes about 36% of pyritic + organic sulfur removal due to coupled mechanisms (cavitation + streaming). In general, ultrasonic method is a time-dependant process. To intensify the ultrasonic treatment method, some insights into its kinetics is needed. Suitable reagents are to be used in order to minimize the treatment time and reagent consumption, and to maximize sulfur removal.

4.4.3.3 Summary

- The cavitation-dominant frequency characterizing the 20 kHz probe and 25 kHz tank systems produces fine coal particles, which enable intimate contact with the strong oxidizing agents produced by the ultrasonic system, leading to efficient coal desulfurization.
- The streaming-dominant frequency present in the 430 kHz tank system removes sulfur from coal only by leaching of impurities induced by streaming effect. The radicals produced by the ultrasound penetrate the pores of the coal particles due to micro-streaming effect.
- In almost all cases, sulphate-sulfur removal is more than 90%, as observed in aqueous-based ultrasonic desulfurization.
- Ultrasonic aqueous-based de-sulfurization seems to be adequate for dealing with high-sulfur coal, which is composed mainly of sulphate sulfur.

4.5 Results for Solvent-Based Coal Beneficiation

4.5.1 Introduction

In this study, state-of-the-art ultrasonic equipment, spanning the frequency range from highly-cavitational (<100 kHz), to intermediate (100–200 kHz) to mostly acoustic-streaming dominated (>470 kHz), were employed to conduct washing trials of high-ash Indian coals from three mines (Belphar, Dipka and Talcher), and in three pre-processed conditions—run-of-the-mine (ROM), conventionally-washed, and reject. Size reduction characteristics were quantified by sieve analysis, and ash content was determined by measuring the residue left after burning coal in a muffle furnace. A Taguchi L_{27} fractional-factorial matrix was designed to assess the individual effects of key process variables—ultrasonic frequency and power, sonication time, coal-to-solvent (Methanol) ratio and feed size of coal particle, ultrasonic wash temperature, and source and condition of the coal. In this study, methanol is used as solvent. The objective of this work is to evaluate sono-washing of coal as a method to effectively reduce coal impurities (including ash) while maintaining a certain minimum size.

4.5.2 De-Ashing

High level analysis (HLA) is performed by averaging measured data for each level of a single parameter, then plotting the averaged data against all the levels of that parameter. In our experiment, a total 162 data points measured can be conveniently analyzed in this manner and is shown in Appendix II. It should be pointed out that following traditional Taguchi DOE methodology, effects of all other controllable parameters have been averaged over the entire DOE; this enables isolation of effects due to a single variable.

4.5.2.1 Effect of Ultrasonic Frequency on Size Reduction and Ash Removal

Size reduction and contaminant-leaching are time-dependent phenomena. The frequency and amplitude of the ultrasonic field have a direct influence on coal particle size reduction and ash-removal characteristics. Lower frequency and higher amplitude, both of which result in increased cavitation intensity, lead to greater particle breakage. Thus, as ultrasonic frequency increases, there is a monotonic decrease in size reduction of coal was observed from Fig. 4.13. For the Talcher coal, 132 kHz appears to be optimal for ash removal, the reason being that both the mechanisms are present in equal measure. The ash removal occurs by the combined mechanism of cavitational breakage and acoustic leaching.

4.5.2.2 Effect of Initial Size of Coal on Size Reduction and Ash Removal

The initial size of the coal particle also affects the size reduction and ash removal. Three different size ranges of coal particles: (− 1 + 0.09), (− 2 + 1) and (− 4 + 2) mm were used for this experimental investigation. From Fig. 4.14, initial coal particle size has a non-monotonic effect on size reduction efficiency, with a minimum occurring in an intermediate size range. However, the variation is small due to less sonication time for all the experiments. The ash removal efficiency for the coals tested increases with decreasing initial size, suggesting that acceleration and impact of suspended particles in the acoustically-coupled medium is the key determinant in causing this effect.

4.5.2.3 Effect of Coal: Solvent (Methanol) Ratio on Size Reduction and Ash Removal

Methanol was used as a solvent for the ultrasonic coal washing experiment. Three different coal-solvent ratios were used: 1:2, 1:4 and 1:6 volume %. Figure 4.15 shows the effect of coal-solvent ratio on size reduction and ash removal. The % of size reduction decreases with an increasing coal-solvent ratio. In this case, a lower mass-loading of coal assists in both size reduction and impurity removal. As the solvent volume increases for a fixed coal volume, both size reduction and the % of ash removal increase. However, the change is small in the range of parameter investigated.

4.5.2.4 Effect of Initial Temperature on Size Reduction and Ash Removal

Temperature has a significant effect on the cavitation phenomenon, which, in a liquid medium, is affected by its surface tension, viscosity and vapor pressure. Increasing temperature results in a reduction in the acoustic cavitation threshold, increase in vapor pressure of the liquid, decrease in surface tension, and reduction in the viscosity of the liquid medium. The decrease in viscosity decreases the magnitude of the natural cohesive forces acting on the liquid, and thus, decreases the cavitation threshold. Lower cavitation thresholds translate into ease of cavity formation, thereby making room temperature more favorable for particle breakage. Three different initial temperatures were used to model the influence on size reduction: namely, 25, 35 and 40 °C. Figure 4.16 shows the effect of temperature on size reduction and ash removal. It has been observed that room temperature appears to be favorable for size reduction and associated ash removal. The effect, however, is rather small especially for size reduction.

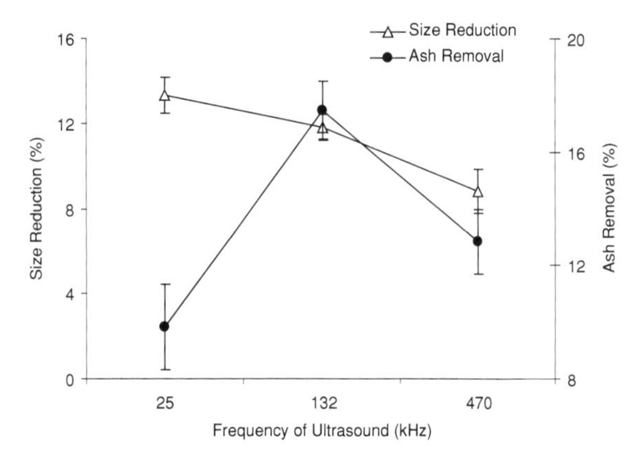

Fig. 4.13 Effect of ultrasonic frequency on size reduction and ash removal for Talcher coal

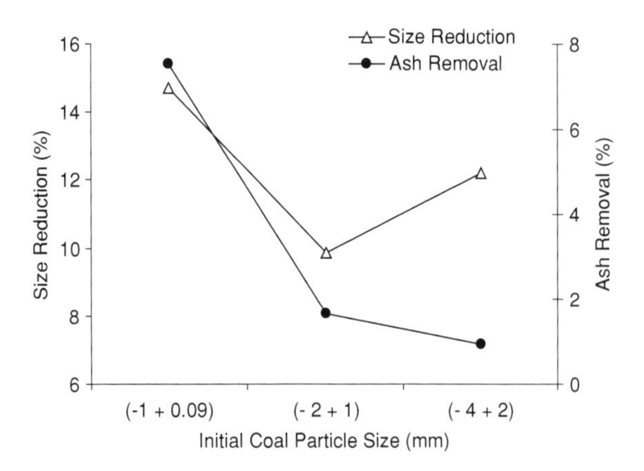

Fig. 4.14 Effect of coal particle feed size on size reduction and ash removal for Belphar coal

4.5.2.5 Effect of Coal Pre-Conditioning on Size Reduction and Ash Removal

Figure 4.17 shows the % of size reduction and ash removal for coals in various processing states (ROM, Reject and Washed) received from various mines. It can be observed that % ash removal is larger for conventionally washed coal, the reason being the higher surface moisture content resulting from hydroscopic moisture absorption during coal mining and conventional washing, which diminishes the adhesive/cohesive strength of the particle. This suggests that an optimum strategy for incorporating ultrasonic wash would be prior to normal wash, not subsequent to. In such an arrangement, ultrasonic wash will loosen ash and other impurities from the coal particles, which would subsequently be rinsed away in the normal wash. The size reduction appears to be rather insensitive to the pre-conditioning of the coal.

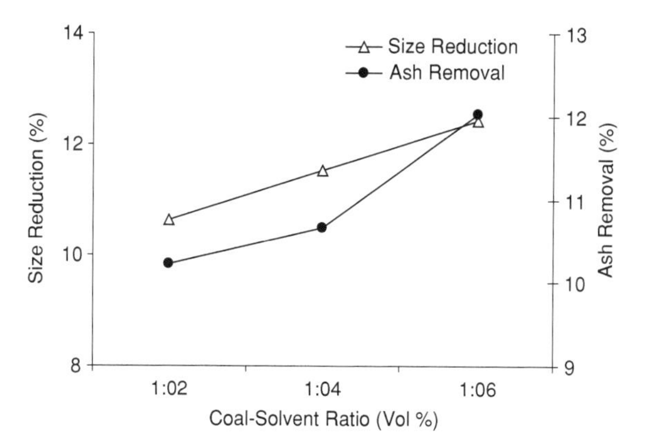

Fig. 4.15 Effect of coal-to-solvent ratio on size reduction and ash removal for Dipka coal

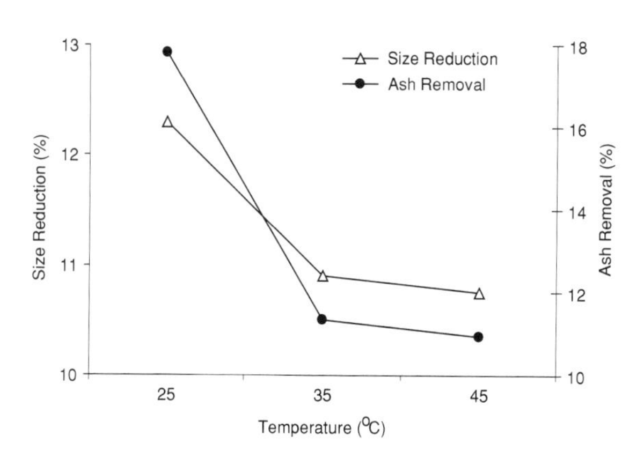

Fig. 4.16 Effect of temperature on size reduction and ash removal for Talcher coal

4.5.2.6 Effect of Three Different Mining Location Coal on Size Reduction and Ash Removal

Figures 4.18 and 4.19 shows variation of size reduction and ash removal for Belphar, Dipka and Talcher coals with three different pre-processing states. From Fig. 4.18, it may been observed that although the variation is small, the size reduction is highest for Dipka washed and Dipka run of mine (ROM) coal, possibly because of the morphology and properties of the coal particle, such as softness, brittleness, and presence of micro-pores or micro-fractures on the surface. From a size reduction viewpoint, all three coals tested show the same qualitative trends, i.e., washed coals shows highest size reduction, next are ROM coals, with reject coals showing a minimum for all three mining locations. Higher surface moisture content in washed coal softens the coal material in comparison with the other pre-processing states (ROM and Reject coal). Figure 4.19 shows the variation of ash removal for coals from the three mining location coals. It indicates that there is a close relation between size reduction and ash removal. Washed coals also show highest ash removal for all three mining locations, followed by ROM coals. However, the reject coals show very little ash removal capability.

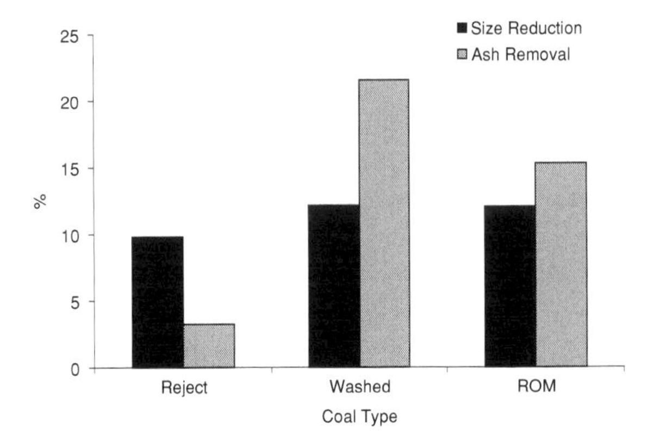

Fig. 4.17 Effect of coal preparation on size reduction and ash removal for Talcher coal

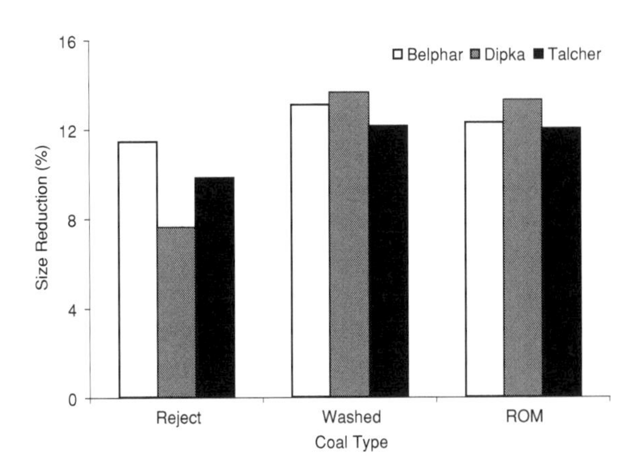

Fig. 4.18 Variation of size reduction with three mining location coals for different pre-processing states

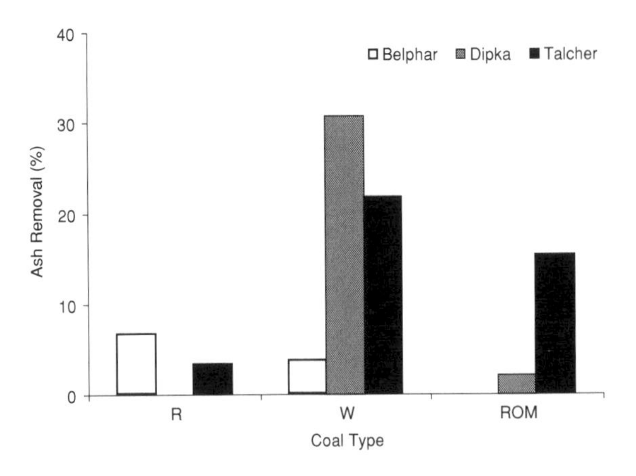

Fig. 4.19 Variation of ash removal with three mining location coals for different pre-processing states

Table 4.7 Optimum value of process parameter for De-Ashing

Parameter	Optimal value
Ultrasonic frequency (kHz)	132
Coal/Methanol ratio (vol%)	1:6
Temperature (°C)	35
Coal condition	–
Initial coal size (mm)	(− 1 + 0.09)

Table 4.8 Percentage of ash reduction at optimum process condition

S.no	Coal type	% of Ash reduced
1	Belphar rejects	11.62
2	Belphar washed	22.46
3	Belphar ROM	8.56
4	Dipka rejects	1.32
5	Dipka washed	21.41
6	Dipka ROM	52.78
7	Talcher rejects	16.59
8	Talcher washed	35.92
9	Talcher ROM	27.74

Based on these DOE results, an optimum set of process conditions has been identified that results in minimum size reduction accompanied by maximum ash removal. These are listed in Table 4.7.

In order to validate the above set of parametric conditions as optimal for ash removal, an independent, fresh sampling of coal from each mine was subjected to ultrasonic washing under these process conditions, and the results were compared to the corresponding un-sonicated coal. Results are summarized in Table 4.8. It has been observed that the average ash removal by optimized ultrasonic wash is of the order of 20%. This validation trial reconfirms the earlier findings that under an optimized set of ultrasonic conditions, the coal can be made quantifiably cleaner in terms of heavy as well as light impurities.

4.5.2.7 Summary

- As ultrasonic frequency increases, a monotonic decrease in size reduction of coal was observed. For the Talcher coal, 132 kHz appears to be optimal for ash removal. The ash removal occurs by the combined mechanism of cavitational breakage and acoustic leaching.
- In general, the size reduction efficiency is rather insensitive to a number of process parameters including the initial particle size, mass loading of coal,

moisture content, etc. The ash removal efficiency for all three coals tested increases with decreasing initial size, suggesting that acceleration and impact of suspended particles in the acoustically-coupled medium is the key determinant in causing this effect.

- In general, a higher mass-loading of coal assists in impurity removal for aqueous ultrasonic wash; however, a lower volume concentration of coal appears to be preferable for solvent ultrasonic wash.
- It has been observed that for ash removal, a temperature slightly higher than ambient appears to be optimal, especially for Belphar coal. The leaching process is partly chemical in the presence of a solvent, and hence is accelerated at higher temperatures.
- Size reduction is higher for washed coal, the reason being the higher surface moisture content resulting from hydroscopic moisture absorption during coal mining and conventional washing, which diminishes the adhesive/cohesive strength of the particle.
- percentage of ash removal was higher for Talcher coal, and least for Belphar coal.
- Washed coals also show highest ash removal.
- The high moisture content in ash and coal leads to increased softness and cohesiveness, resulting in low ash removal.
- An optimum strategy for incorporating ultrasonic wash would be prior to normal wash, not subsequent to. In such an arrangement, ultrasonic wash will loosen ash and other impurities from the coal particles, which would subsequently be rinsed away in the normal wash.

4.5.3 De-Sulfurization Studies

In order to achieve minimum treatment time, less reagent consumption and maximum removal, reagent-based ultrasonic desulfurization was investigated. Pyritic and organic sulfur show high resistance to ultrasonic treatment in aqueous medium. To intensify the ultrasonic treatment method, 2N nitric acid and 3-volume percentage of hydrogen peroxide were used as reagents. The reasons for choosing the above mentioned reagent concentrations are that as per IS1350 (forms of sulfur determination) procedure, 2N of nitric acid is used to extract entire inorganic sulfur from coal by 30 min of boiling, and higher volume percent of H_2O_2 resulted in foaming and uncontrolled reactions.

4.5.3.1 Comparison Between Conventional and Ultrasonic Method

To assess the enhancement effect of the ultrasonic method, conventional methods of desulfurization were compared. 2N of HNO_3 and 3-volume % of H_2O_2 were used as reagents. 10 g of 212 μm sieve pass-through high-sulfur lignite coal I and 100 ml of reagents were taken in a 250 ml beaker. The same coal-to-reagent ratio was maintained for reagent-based soaking, stirring and sonication.

Figure 4.20 shows a comparison between conventional and ultrasonic methods of coal de-sulfurization. 120 h soaking of high-sulfur coal I in reagents leads to about 46% removal by HNO_3 and 35% removal by H_2O_2, due mainly to chemical reaction. The majority of sulfur removal occurs only by surface reaction. The reaction moves gradually towards core of the particle as time progresses, thereby extending the process. Stirring was conducted for about 1 h at 1,000 rpm. It is apparent that using nitric acid at 1,000 rpm, 29% removal is possible; in the case of H_2O_2, 27% removal was observed. In order to assess the effect of ultrasound on reagent-based coal desulfurization, 20 kHz probe is used at 500 W input power. 30 min of sonication leads to 74% of TSR by H_2O_2 and 23 min of sonication leads to 87% of TSR by HNO_3.

4.5.3.2 Effect of Sonication on TSR and Reaction Mixture Temperature

The reaction mixture was sonicated at intervals of 10, 20 and 30 min and the corresponding temperature rise of the reaction mixture was monitored using data-logger. The prime reason for measuring mixture temperature is to track the rate at which the reaction proceeds during ultrasound irradiation. In Fig. 4.21, for 3-volume percentage of H_2O_2, the percentage removal of sulfur is seen to increase with increasing sonication time. Interestingly, even the low reagent concentration (3 volume %) and the short period of sonication (10 min) lead to more than 63% removal. The reason is that the directional probe creates an intense ultrasonic field near the top which leads to rapid breakage of the coal particles, thereby accelerating the reaction between H_2O_2 and sulfur. For both the reagents, the kinetics is very rapid, when using 3% H_2O_2, 30 min sonication leads to 74% of total sulfur removal, for 2N HNO_3, 23 min of sonication leads to 87% removal. The slope of the curve at 3% hydrogen peroxide condition is steeper than 2N nitric acid condition.

The reaction between hydrogen peroxide and sulfur is exothermic; this is confirmed by a significant temperature rise in the reaction mixture within a short span of time as shown in Fig. 4.22. Once the maximum temperature is reached, the loss of heat by evaporation (the experiments were conducted in an open atmosphere) more than accumulated for the generation of heat resulting from further removal of sulfur. As a result, the temperature actually decreases with aqueous sulfur removal, the reaction is not intensified greatly in the beginning enough to produce a peak in the temperature, the mixture temperature reaches a high value of 65 °C asymptotically. Similarly, the temperature rise during reagent soaking and reagent-based stirring are also fairly small. The temperature rise during 30 min of solvent soaking and stirring is only 4–8 °C.

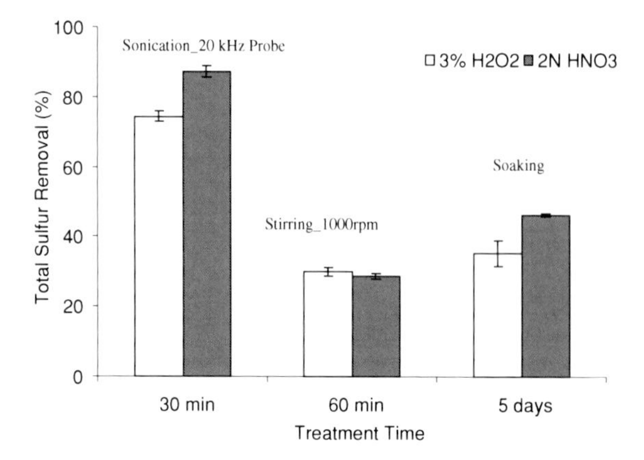

Fig. 4.20 Comparisons between conventional and ultrasonic method

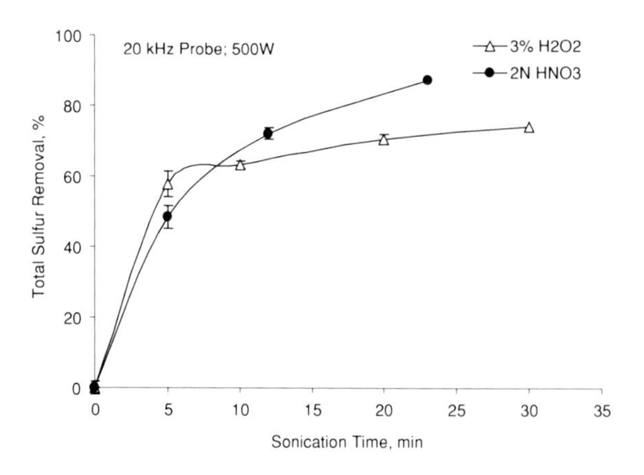

Fig. 4.21 Effect of reagent-based ultrasonic methods on total sulfur removal

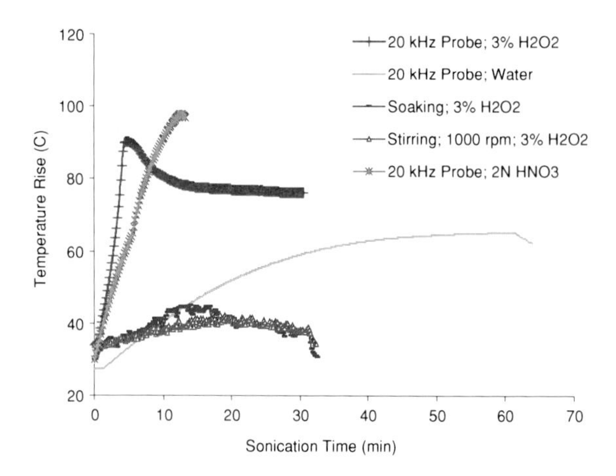

Fig. 4.22 Effect of sonication, soaking and stirring on coal-reagent mixture temperature rise

The trend with 2N HNO_3 is similar to that with H_2O_2. From Fig. 4.22, it may be observed that the temperature rise in nitric acid and coal mixture is higher than in the H_2O_2 case. The maximum temperature achieved during sonication was 98 °C within 12 min, corresponding to 72% sulfur removal.

The reagent-based ultrasonic desulfurization is much faster than aqueous-based or conventional methods of desulfurization. Hydrogen peroxide seems to be a more suitable reagent for removing sulfur from coal than HNO_3 because it is less harmful to the environment and appears to be effective even with 3% by volume concentration whereas typical application of H_2O_2 as an oxidizing agent is up to 6% concentration.

4.5.3.3 Summary

- Reagent-based ultrasonic desulfurization results in higher removal within a short period of treatment time, compared to conventional methods.
- Sonication accelerates the reagent-based reaction; this has been confirmed by increased removal of sulfur and significant temperature rise in the reaction mixture within a short span of time.
- In an ultrasonic field, the reaction between H_2O_2 and sulfur is very fast, and the temperature rise in the reaction mixture within 4 min of treatment is 90 °C. Due to fast reaction and high sulfur removal, low reagent concentration and lack of by-products, hydrogen peroxide seems to be the most suitable reagent for ultrasonic desulfurization.
- Hydrogen peroxide removes all forms of sulfur in an ultrasonic field within short span of treatment time.

In view of this, systematic experiments have been conducted on reagent-based coal de-sulfurization. These are described in the next chapter.

References

1. Mason TJ, Lorimer JP, Bates DM (1992) Quantifying sonochemistry: casting some light on a black art. Ultrasonics 30:40–42
2. Toma M, Fukutomi S, Asakura Y, Koda S (2011) A calorimetric study of energy conversion efficiency of a sonochemical reactor at 500 kHz for organic solvents. Ultrason sonochem 18:197–208
3. Luche JL (1994) Effect of ultrasound on heterogeneous system. Ultrason Sonochem 1:111–118
4. Ratoarinoro N, Contamlne F, Wilhelm AM, Berlan J, Delmas H (1995) Activation of a solid-liquid chemical reaction by ultrasound. Chem Eng Sci 50:554–558
5. Webster E (1963) Cavitation. Ultrasonics 1:39–48
6. Gopi KR, Nagarajan R (2008) Advances in nano-alumina ceramic particle fabrication using sono-fragmentation. IEEE Trans Nanotechnol 7:532–537

Chapter 5
Experimental Studies and Mechanistic Modeling of Reagent-Based Ultrasonic Coal De-Sulfurization

5.1 Motivation

Conventional reagent-based de-sulfurization removes all forms of sulfur, but at the expense of reagent volume, reagent concentration and treatment time. At the end of the treatment, there will be a problem of neutralizing the treated coal sample. Probe-type ultrasonics is localized, and the non-uniformity in energy transmission renders unsuitable for large-scale coal processing. These reasons have led to the investigation of reagent-based coal de-sulfurization using tank-type ultrasonic system. Three different ultrasonic frequencies were used for this experimental investigation: 25 kHz (purely cavitation-dominant), 430 kHz (streaming-dominant) and 58/192 kHz (combination of both mechanisms). These ultrasonic tanks were operated at 500 W input power (including the dual system, where low and high frequency-coupled ultrasonics is used; each frequency generator operates at 500 W power, and drives an alternating diagonal set of transducers). The study employed a Taguchi fractional-factorial L_{27} DOE. The effects of ultrasonic frequency, reagent concentration, sonication time, coal particle size and coal concentration, and reagent volume on reagent-based ultrasonic coal-desulfurization are presented here.

5.2 Experimental

5.2.1 Materials

The coal to be studied was received from Giral mine, Rajasthan, India. The coal sample was air dried, and ground using ball mill. Then, the coal sample was sieved using Indian Standard sieves to required particle size as per DOE. The proximate and sulfur analysis of the coal samples were performed according to ASTM Standard procedures and are shown in Table 4.1.

B. Ambedkar, *Ultrasonic Coal-Wash for De-Ashing and De-Sulfurization*, Springer Theses, DOI: 10.1007/978-3-642-25017-0_5,

Table 5.1 Experimental factors and levels for de-sulfurization experiments

S. No	Parameters	Level 1	Level 2	Level 3
1	Frequency (kHz)	25	58 + 192	430
2	Reagent concentration (mol 1-1 or Volume %)	1 mol 1-1 0.5 mol 1-1 1.5%	3 mol 1-1 1 mol 1-1 3%	5 mol 1-1 2 mol 1-1 6%
3	Sonication time (min)	10	20	30
4	Avg. particle size (μm)	106	406	800
5	Reagent volume (ml)	30	60	100

5.2.2 Experimental Design

Taguchi's parameter design provides a simple and systematic approach for optimization of design for quality, performance and cost. Table 5.1 shows the factors and levels for L_{27} DOE. To investigate reagent-based ultrasonic coal desulfurization with five controllable three-level factors, L_{27} array including 27 tests was chosen and shown in Table 5.2. L_{27} DOE is repeated for three different reagents.

5.2.3 Experimental Procedure

As-received coal was ground to required size and air-dried. The proximate analysis of the coal samples was performed using IS 1350 procedure. De-sulfurization experiments were carried out by ultrasonic methods. Three reagents (HCl, HNO_3 and H_2O_2) were used. The reagents used were of analytical grade. 10 g of coal of known size range were taken in a 150 ml beaker, and reagent was poured into the beaker, per the trial condition specified in DOE (Table 5.2). The mixture was subjected to sonication. Then, the treated sample mixture was filtered, washed with water and dried for about 5 h in an oven at 105 °C. The treated coal sample was analyzed for sulfur content according to the procedure given in IS 1350.

5.2.4 Reaction Mechanism

Hydrochloric acid reacts with elemental sulfur as well as with pyritic sulfur to give hydrogen sulfide in gaseous forms; it is found to be inefficient with respect to organic sulfur removal.

$$S + 2HCl \rightarrow H_2S + Cl_2$$

$$FeS_2 + 2HCl \rightarrow H_2S + FeCl_2 + S$$

Table 5.2 L_{27} orthogonal array for de-sulfurization experiments

Trial No.	Frequency kHz	Reagent concentration			Time min	Coal size micron	Solvent volume ml
		HCl N	HNO_3 N	H_2O_2 Vol%			
1	25	1	0.5	1.5	10	106	30
2	25	1	0.5	1.5	20	406	60
3	25	1	0.5	1.5	30	800	100
4	25	3	1	3	10	106	30
5	25	3	1	3	20	406	60
6	25	3	1	3	30	800	100
7	25	5	2	6	10	106	30
8	25	5	2	6	20	406	60
9	25	5	2	6	30	800	100
10	Dual	1	0.5	1.5	10	406	100
11	Dual	1	0.5	1.5	20	800	30
12	Dual	1	0.5	1.5	30	106	60
13	Dual	3	1	3	10	406	100
14	Dual	3	1	3	20	800	30
15	Dual	3	1	3	30	106	60
16	Dual	5	2	6	10	406	100
17	Dual	5	2	6	20	800	30
18	Dual	5	2	6	30	106	60
19	430	1	0.5	1.5	10	406	60
20	430	1	0.5	1.5	20	106	100
21	430	1	0.5	1.5	30	406	30
22	430	3	1	3	10	800	60
23	430	3	1	3	20	106	100
24	430	3	1	3	30	406	30
25	430	5	2	6	10	800	60
26	430	5	2	6	20	106	100
27	430	5	2	6	30	406	30

Nitric acid is a good reagent for the total inorganic sulfur. It primarily reacts with pyritic sulfur and converts it to water-soluble ferrous sulphates. A main disadvantage is that it produces sulphuric acid as by-product, and is inefficient with respect to organic sulfur removal.

$$3FeS_2 + 18HNO_3 = Fe(So_4)_3 + Fe(No_3)_3 + 3H_2SO_4 + 15NO + 6H_2O$$
$$3FeS_2 + 18HNO_3 = Fe(So_4)_3 + H_2SO_4 + 10NO + 4H_2O$$

Hydrogen peroxide reacts with all forms of sulfur. The reaction mechanism is given below. It converts all forms of sulfur into water-soluble sulphates without producing any harmful by-products.

$$H_2S + H_2O_2 \rightarrow S + 2H_2O$$
$$HS^- + 4H_2O_2 \rightarrow SO_4^{2-} + 4H_2O + H^+$$
$$S^{2-} + 4H_2O_2 \rightarrow SO_4^{2-} + 4H_2O$$
$$2S_2O_3^{2-} + H_2O_2 \rightarrow S_4O_6^{2-} + 2OH^-$$
$$S_4O_6^{2-} + 3H_2O_2 \rightarrow S_3O_6^{2-} + SO_4^{2-} + 2H_2O + 2H^+$$
$$S_3O_6^{2-} + H_2O_2 + H_2O \rightarrow 3SO_3^{2-} + 4H^+$$
$$SO_3^{2-} + H_2O_2 \rightarrow SO_4^{2-} + H_2O$$
$$2RSH + H_2O_2 \rightarrow RSSR + 2H_2O$$
$$RSSR + 5H_2O_2 + 2OH^- \rightarrow 2RSO_3^- + 6H_2O$$

5.3 Results and Discussion

High Level Analysis has high statistical significance. HLA is performed by averaging measured data for each level of a single parameter, then plotting the averaged data against all the levels of that parameter. A total 162 data point of desulfurization experiments were measured and are shown in Appendix III. The effect of the individual parameters is analyzed below.

5.3.1 Effect of Ultrasonic Frequency on Total Sulfur Removal

Figure 5.1 shows effect of ultrasonic frequency on total sulfur removal (TSR) from coal. In a 25 kHz ultrasonic frequency-field, cavitation causes particle breakage thereby producing fine particles. Fines produced by the ultrasound have high surface area and high sphericity. These characteristics of the coal particle lead to enhanced chemical reaction by promoting intimate contact between the particle and reagents. Violent collapse of bubbles creates very high turbulence in the ultrasonic field, which causes uniform mixing throughout the reaction mixture.

Figure 4.2 shows fluid behavior at various ultrasonic frequencies. At 25 kHz, the liquid medium looks stagnant, but is experiencing very high intense cavitation in the tank (Fig. 4.2a). The stagnancy of the liquid is evidence that the 25 kHz frequency is only cavitation-dominant. Dual-frequency is the combination of low and high frequency of ultrasound. Low frequency of ultrasound causes particle breakage, and high frequency of ultrasound causes leaching effect due to acoustic streaming. The relatively high removal of total sulfur attained in dual-frequency is thus due to this combined mechanism. The liquid behavior in the dual-frequency tank is somewhat different, resembling a slithering movement (Fig. 4.2b). In almost all cases, dual-frequency renders highest sulfur removal due to combined

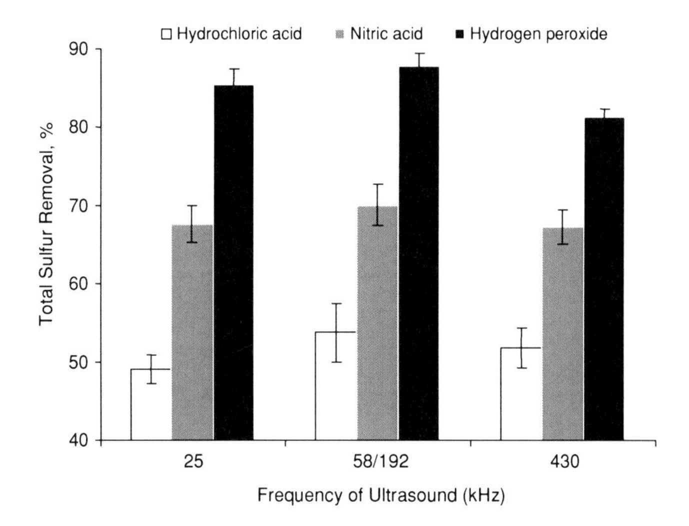

Fig. 5.1 Effect of ultrasonic frequency on TSR

mechanisms present in equal measure. In 430 kHz tank, acoustic streaming is the dominant phenomenon. This has been confirmed by "fountain" effect seen in the center of the tank (Fig. 4.2c). Lack of size-reduction at 430 kHz has been confirmed by size analysis of coal particle before and after sonication at various frequencies (Fig. 4.6d). Hence, removal occurs mainly because of streaming phenomena.

Comparing with Fig. 4.22, it can be seen that in an ultrasonic bath, H_2O_2 is proving to be a better reagent than HNO_3 for TSR. This may be due to the difference in the ultrasonic intensity levels in a directional probe (as used in Fig. 4.22) and an ultrasonic bath (as used in Fig. 5.1). Since H_2O_2 is more environmentally friendly and since an ultrasonic bath is more feasible on an industrial scale, these results point to H_2O_2 as the preferred reagent of ultrasonic-based de-sulfurization.

5.3.2 Effect of Reagent Concentration on Total Sulfur Removal

Figure 5.2 shows effect of reagent concentration on reagent-based ultrasonic coal desulfurization. In Case 1, 1, 3 and 5 N concentration of HCl was used for this investigation. From Fig. 5.2, 56% total sulfur was removed using 5 N of HCl; using lower HCl concentration (1 and 3 N), the maximum S removal was about 50%. In Case 2, three different concentrations of nitric acid were used for the investigation, i.e., 0.5, 1, and 2 N. Total sulfur removal increases with increasing reagent concentration. From Fig. 5.2, it has been observed that 75% of sulfur is removed at high reagent concentration. Interestingly, 63% of total sulfur removal was achieved with the lowest reagent concentration (0.5 N HNO_3). In Case 3, 1.5, 3 and 6 volume % of H_2O_2 concentrations were used. Figure 5.2 shows that sulfur

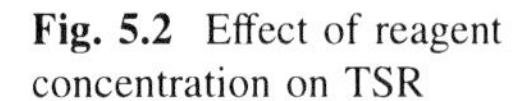

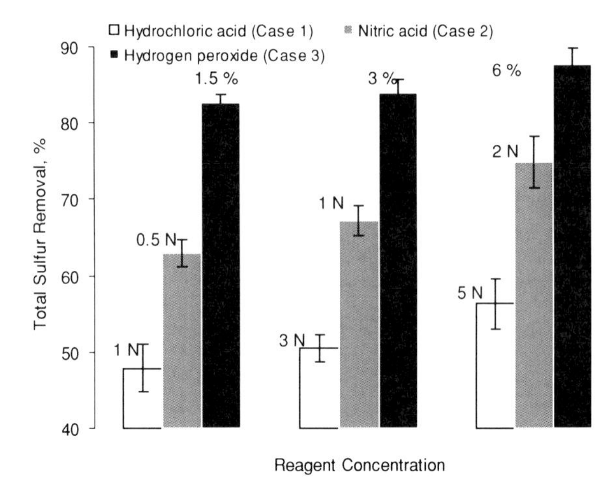

Fig. 5.2 Effect of reagent concentration on TSR

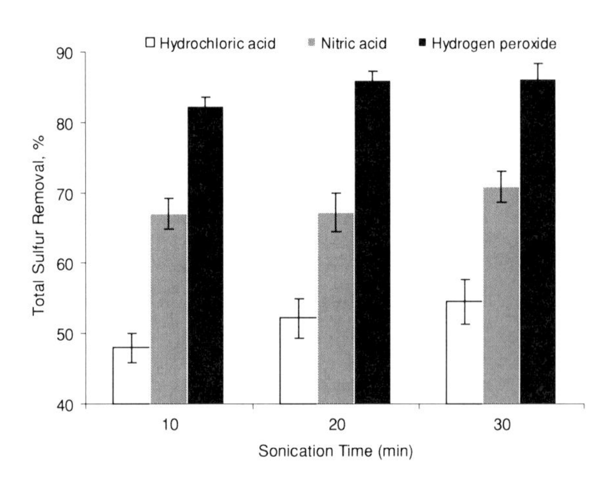

Fig. 5.3 Effect of sonication time on TSR

removal increases with increasing hydrogen peroxide concentration. Using 6 volume % of hydrogen peroxide results in 88% of total sulfur removal, but 80% of total sulfur removal was achieved with just 1.5 volume % of H_2O_2. All the cases show a nearly linear dependence on reagent concentration.

5.3.3 *Effect of Sonication Time on Total Sulfur Removal*

The main purpose of using ultrasonic methods in coal desulfurization is to alter the reaction pathways and to shorten the reaction time. Figure 5.3 shows the effect of sonication time on total sulfur removal. The coal sample mixture was sonicated for three different time intervals, i.e., 10, 20 and 30 min. The removal of total sulfur increases with increasing time of sonication. In the case of HCl, 55% of total sulfur

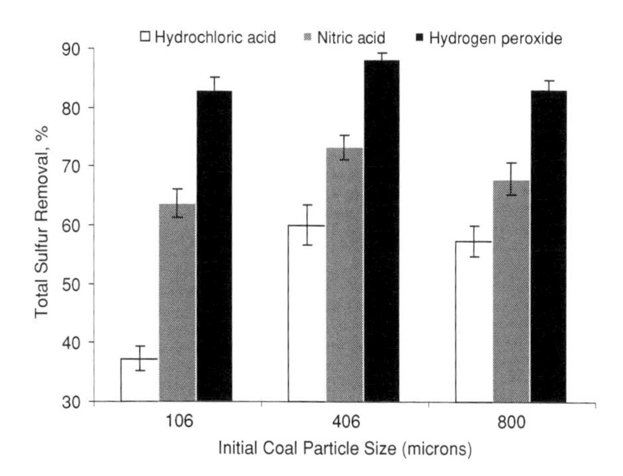

Fig. 5.4 Effect of coal particle size on TSR

was removed in 30 min of sonication, for HNO_3 71% of total sulfur and finally for H_2O_2, 86% of total sulfur was removed at maximum time of sonication. In the case of hydrogen peroxide, the initial rate of TSR is high compared to other two cases. This shows that hydrogen peroxide is more reactive in an ultrasonic field compared to the other two.

5.3.4 Effect of Coal Particle Size on Total Sulfur Removal

Particle size plays a major role in ultrasonic coal desulfurization. Three different sizes of coal particle were chosen (i.e., 106, 406 and 800 μm) to study the effect on coal desulfurization. Figure 5.4 shows the effect of coal particle size on total sulfur removal. In all cases, the physical effects of ultrasonics, cavitation and streaming, play an interactive role with prevailing chemical effects, albeit with size-dependent differences. It can be seen that for H_2O_2, the effect of size is not very pronounced. It may be expected that TSR would be highest for the smallest size sample; surprisingly, it is least. This may be due to the complicated interaction that takes place between agitation, breakage, attrition and turbulent dispersion. It is possible that particle–particle interaction is less for small particles and for the particle size is such that turbulence suppression takes place. Further investigation of this effect is necessary.

5.3.5 Effect of Reagent Volume on Total Sulfur Removal

Figure 5.5 shows the effect of reagent volume on total sulfur removal. The main objective of reagent-based ultrasonic coal desulfurization is to minimize the amount of reagent utilized for coal de-sulfurization. In this context, three different

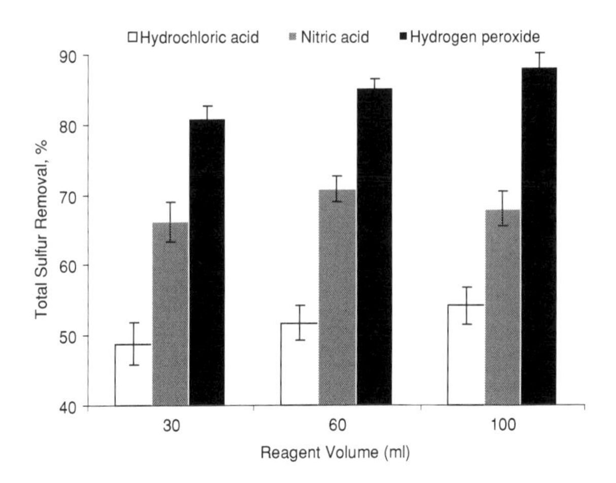

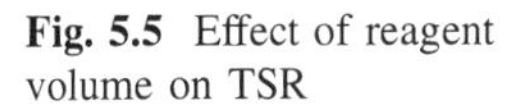
Fig. 5.5 Effect of reagent volume on TSR

volumes of reagent, i.e., 30, 60 and 100 ml, were experimented with. The % removal of total sulfur increases with increasing amount of reagent available for coal desulfurization. These trends were observed in the case of HCl as well as H_2O_2 reagents. Reagent-based ultrasonic methods remove more than 90% of maximum total sulfur even at the lowest volumes included in this investigation. But, nitric acid behaves in a different manner. Highest removal was observed in 60 ml of reagent volume, and it is less for other volumes in the trial. This behavior is related to physical properties of the acoustic field and the reagent. Since cavitation intensity is a volumetric phenomenon, energy per unit volume of reagent is reduced as volume is increased, keeping input power constant. While chemically, more reagents might result in greater sulfur extraction; this is in conflict with the physical acoustic-intensity effect.

Table 5.3 shows density and viscosity of reagents that were considered for this investigation. In general, cavitation and streaming is more pronounced in low-viscosity and low-density fluids. A high-viscosity reagent requires input of a higher threshold energy to initiate cavitation. Once initiated, the cavitation effect is stronger in a high-viscosity fluid. Nitric acid has higher viscosity and density compared to the other two reagents. Initially, desulfurization efficiency increases with reagent volume, since energy and sonication time needed are less; later, it decreases with higher volume of reagent as initiation of cavitation requires that a high-threshold energy barrier be overcome.

5.3.6 Statistical Validation of Results

T-test analysis was performed for assessing statistical significance of the data. This test may be used to determine which parameter has significant effect on total sulfur removal. Critical value of t was found from the T table (www.sjsu.edu). Table 5.4

Table 5.3 Physical properties of reagents

	HCl	HNO_3	H_2O_2
Density (kg/m^3)	1,180	1,510	1,460
Viscosity (kg/m.s) × 10^3	1.9	2.5	1.245

Table 5.4 T statistical value and confidence level of parameters for different reagents

	HCl		HNO_3		H_2O_2	
	Tstat	Confidence (%)	Tstat	Confidence (%)	Tstat	Confidence (%)
Ultrasonic frequency	0.8163	>50	0.6864	50	2.2746	>95
Reagent concentration	2.1447	>95	2.7518	>95	1.9736	>90
Time	1.6695	>80	0.8651	>60	1.5001	>80
Coal particle size	4.7910	>95	0.8028	>50	0.0566	<50
Reagent volume	13.3972	>95	0.3248	<50	2.7678	>95

Table 5.5 Optimum value of process parameters based on maximum TSR

	Ultrasonic frequency, kHz	Reagent concentration	Sonication time, min	Coal particle size, μm	Reagent volume, ml
HCl	Dual	5 N	30	−600 + 212	100
HNO_3	Dual	2 N	30	−600 + 212	60
H_2O_2	Dual	6 volume %	30	−600 + 212	100

shows the value and the corresponding confidence level of parameters for different reagents. In the case of hydrogen peroxide the confidence level of all process parameters is greater than 80% excluding coal particle size. This indicates that, there is a consistency in TSR using hydrogen peroxide as a reagent. The concentration and volume of the reagent and the ultrasonic frequency have a high confidence level indicating the influence of these parameters on total sulfur removal.

5.4 Optimum Conditions and Validation

Optimum set of process parameters for reagent-based ultrasonic coal desulfurization was determined on the basis of highest TSR, and shown in Table 5.5.

To validate the above optimum conditions, two different types of high-sulfur lignite coals were chosen. The first is the one already used for this investigation, and the second one has slightly higher sulfur and ash than the first. The proximate and sulfur analysis of high sulfur-lignite II coal are shown in Table 4.1.

Table 5.6 Percentage removal of total sulfur and ash under optimum conditions for high sulfur lignite I and II

		Total sulfur removal, %	Ash removal, %
Lignite I	HNO_3	82.5	44.3
	H_2O_2	94.8	63.5
Lignite II	HNO_3	74.5	38.0
	H_2O_2	85.5	66.2

The % removal of total sulfur and ash under optimum conditions for high sulfur lignite coals I and II is shown in Table 5.6. An impressive amount of simultaneous removal of sulfur and ash was obtain using optimum set of process parameters for lignite I. For lignite II, sulfur removal of 85% and ash removal of 66% were obtained with H_2O_2 as the reagent. Therefore, the method looks very promising induced for coal beneficiation.

5.5 Mechanistic Modeling of Ultrasound Assisted Reagent-Based Coal De-Sulfurization

The sonochemical activation of various solid–liquid reactions has been widely studied, but the mechanisms of heterogeneous sonochemistry remain poorly understood. Enhancements due to ultrasound may be attributed to its chemical and mechanical effects. The chemical effects of ultrasound are due to the implosion of micro-bubbles, generating free-radicals with a great tendency for reaction. Mechanical effects are caused by shock waves formed during symmetric cavitation or by micro-jets formed during asymmetric cavitation. The high local turbulences can also improve solid liquid mass transfer.

The factors that are considered for experimental optimization of reagent-based ultrasonic coal de-sulfurization are insufficient to predict the relationship between rate of total sulfur removal and the factors that are influencing reagent-based ultrasonic de-sulfurization of coal. This necessitates the formulation of a mechanism-based model to understand and predict the rate of total sulfur removal. Hence, dimensional analysis was performed to identify the relevant mechanism-based non-dimensional groups.

Ultrasound assisted reagent-based coal de-sulfurization involves three basic mechanisms: enhanced mass transfer with chemical reaction; particle breakage, and enhanced reagent/product transport. The first term in Eq. 5.1 represents the enhanced mass transfer with chemical reaction and associated bulk reaction temperature rise. Cavitation collapse produces micro-jets, which impinge on the surface of the solids producing fines and forming cracks; these cracks develop by subsequent collapse of the bubbles, finally causing particle breakage. The resulting fine particles have high surface area, leading to enhanced total sulfur removal, represented as breakage mechanism in second term. Cavitation collapse will generate shock waves which can cause particle cracking through which the leaching agent can enter the interior of particle by acoustic streaming and leach out the contaminants. Diffusion through pores to the reaction zone will be enhanced by the ultrasonic capillary effect. These transport mechanisms are included in the third term.

The general form of the equation is:-

$$\left(\frac{TSR\ \ d_{Pi}}{t\ K_m}\right) = K\left(\frac{R_X\ t\ T_f}{C_s\ T_o}\right)^a \left(A_{Sp}\left(\frac{M_c}{V_s}\right) d_{Pf}\right)^b \left(\frac{f^2\ D_{eff}}{e_d}\right)^c \tag{5.1}$$

where

TSR = Total sulfur removal (%)
f = Frequency of ultrasound (Hz)
Cs = Initial reagent concentration (mol/L or gm/L)
t = Sonication time (s)
A_{sp} = Total specific surface area (m^2/kg)
R_X = Rate of reaction with mass transfer (L/mol × s)
d_{pi} = Initial coal particle size (microns)
d_{pf} = Final coal particle size (microns)
e_d = Energy dissipated/unit mass of liquid (W/kg)
To = Initial temperature (°C)
T_f = Final temperature (°C)
V_L = Reagent volume (m^3)
M_C = Mass of coal (Kg)
K_m = Mass transfer co-efficient (m/s)
D_{eff} = Effective diffusivity (m^2/s)
Ro = Initial radius of coal particle (microns)
X_A = Fractional conversion of solid reactant
M_A = Molecular weight of sulfur
K = Model constant
a, b, c = Model parameters

Here, the overall (solid–liquid) reaction rate R_X is given by Ratoarinoro et a. [1]

$$R_X = \left(\frac{1}{V_L}\right) K_m A_{SP}\, C_s \tag{5.2}$$

The total specific surface (A_{sp}) area can be expressed function of particle size (d_{pf}). Hence overall reaction rate can be re-written as

$$R_X = \left(\frac{6\, M_p}{\rho_p\, d_{pf}\, V_L}\right) K_m\, C_s \tag{5.3}$$

Where M_P is the mass of the particle

At the end of each experiment, the final mixture temperature (T_f) was measured using mercury thermometer. T_f and the initial temperature are used to account for the effect of temperature in the reaction rate. The breakage mechanism of the coal particle is considered in this model to be strongly influenced by the liquid medium. The reason behind is reagents that are used in this experimental investigation are low viscosity fluids like HCl, HNO_3 and H_2O_2. The physical properties of these dilute solutions closely mimic physical properties of aqueous medium. Hence, effect of fluid viscosity, surface tension and density on coal particle breakage is negligible. Therefore only the volume of the solvent is considered in the model.

The mass transfer coefficient (K_m) is given by Sano et al. [2].

$$K_m = \frac{D_{eff}}{d_{Pf}} \left[2 + 0.4 \left(\frac{e_d \, d_{Pf}^4 \rho_L^3}{\mu_L^3} \right)^{1/4} \left(\frac{\mu_L}{\rho_L D_{eff}} \right)^{1/3} \right] \tag{5.4}$$

Here, the energy dissipated (e_d) per unit mass of liquid is calculated experimentally by using calorimetric study and is given in Table 4.2.

The effective diffusivity (D_{eff}) is calculated from overall SIM (Sharp Interface Model) model [3] for a spherical pellet. This is given by the well known relation

$$\frac{R_o^2}{6 D_{eff}} \left[1 - 3(1 - X_A)^{2/3} + 2(1 - X_A) \right] = \left(\frac{C_S M_A}{2 \rho_A} t \right) \tag{5.5}$$

where X_A, the fractional conversion of the reactant, is obtained from the experiment.

From dimensional parametric-effect analysis, it is apparent that mass transfer with chemical reaction and breakage function are the most sensitive mechanisms in determining reagent-based ultrasonic coal desulfurization, since ultrasonically-enhanced chemical reaction takes place as soon as coal particle breakage occurs as shown in Eqs. 5.6, 5.7 and 5.8. Transport mechanism (streaming and diffusional effect) has a second-order effect.

The limits of each non-dimensional group are shown in Table 5.7. The limits of second non-dimensional group is identical for all the cases, the reason is coal particle breakage in an aqueous medium is included in this model for all the cases.

Table 5.7 Limits of non-dimensional groups

	HCl	HNO_3	H_2O_2
$\frac{R_X \; t \; T_f}{C_s \; T_o}$	3.5×10^2–6.25×10^7	5.82×10^2–1.09×10^8	8.85×10^1–4.47×10^6
$A_{Sp} \left(\frac{M_c}{V_s}\right) d_{Pf}$	1.06–3.5	1.06–3.5	1.06–3.5
$\frac{f^2 D_{eff}}{e_d}$	1.13×10^{-2}–5.58×10^{-7}	1.3×10^{-1}–9.83×10^{-6}	1.4×10^{-2}–6.9×10^{-7}

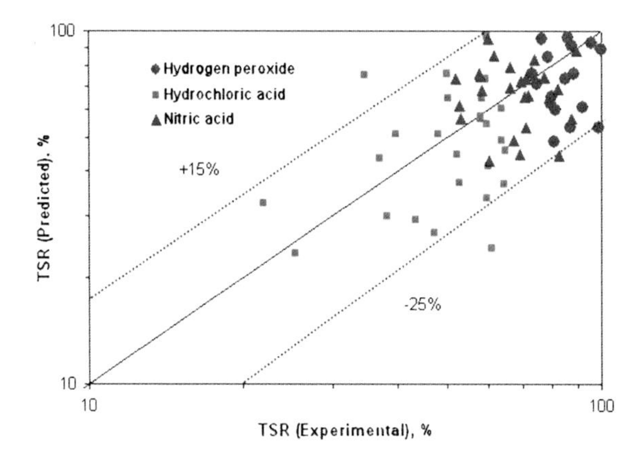

Fig. 5.6 Comparison between experimental and predicted TSR

HCl

$$\left(\frac{TSR\ \ d_{Pi}}{t\ K_m}\right) = 19.88\left[\frac{\left(A_{Sp}\left(\frac{M_c}{V_s}\right)d_{Pf}\right)^{2/5}}{\left(\frac{R_X\ t\ T_f}{C_s\ T_o}\right)^{2/5}\left(\frac{f^2\ D_{eff}}{e_d}\right)^{0.065}}\right] \tag{5.6}$$

HNO_3

$$\left(\frac{TSR\ \ d_{Pi}}{t\ K_m}\right) = 19.88\left[\frac{\left(A_{Sp}\left(\frac{M_c}{V_s}\right)d_{Pf}\right)^{1/2}}{\left(\frac{R_X\ t\ T_f}{C_s\ T_o}\right)^{2/5}\left(\frac{f^2\ D_{eff}}{e_d}\right)^{0.04}}\right] \tag{5.7}$$

H_2O_2

$$\left(\frac{TSR\ \ d_{Pi}}{t\ K_m}\right) = 33.1\left[\frac{\left(A_{Sp}\left(\frac{M_c}{V_s}\right)d_{Pf}\right)^{2/5}}{\left(\frac{R_X\ t\ T_f}{C_s\ T_o}\right)^{1/3}\left(\frac{f^2\ D_{eff}}{e_d}\right)^{0.069}}\right] \tag{5.8}$$

Figure 5.6 shows the comparison between experimental values of total sulfur removal, and values predicted by the model represented in Eqs. 5.6, 5.7 and 5.8. The model appears to be able to predict the total sulfur removal quite accurately, with reasonable level of confidence.

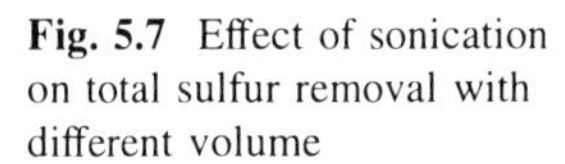

Fig. 5.7 Effect of sonication on total sulfur removal with different volume

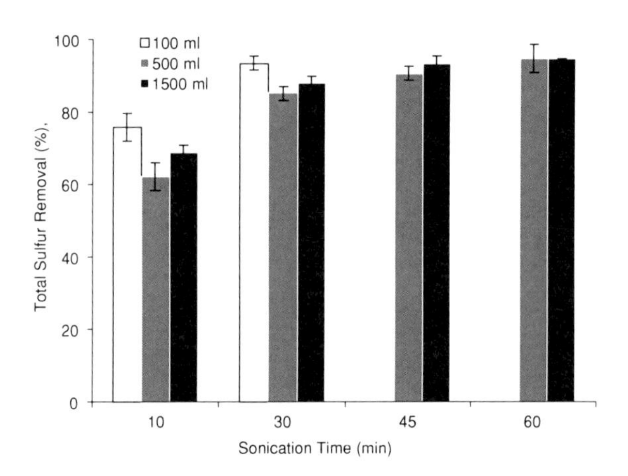

5.6 Scale-Up of Ultrasonic Reagent-Based Coal De-Sulfurization

The main objective of scaling up of ultrasonic method for coal-wash is to achieve better sulfur removal with minimum reagent usage, lowest reagent concentration and high throughput of coal processing. Figure 5.7 shows effect of sonication time on total sulfur removal with different volume of coal-reagent mixture. Three different volumes of coal–reagent mixture (10 g in100 ml, 50 g in 500 ml and 150 g in1500 ml) were taken for this investigation regarding how scaling up of the process effects TSR. From Fig. 5.7, it may be observed that the rate of sulfur removal for first 10 min of sonication is high. It is very close to 80% for 100 ml case and 62–70% for 500 and 1,500 ml cases. Then, the next 20 min of sonication yields a further 20–25% of total sulfur removal. Further 30 min of sonication yields 5–7% of sulfur removal. In almost all the cases, the first 30 min of sonication yields >85% of total sulfur removal. These results show that the sulfur removal results can be scaled up.

An assessment can be made of the amount of coal that can be processed assuming that the experimental results can be done continuously. These estimates are summarized in Table 5.8. The size of lab-scale ultrasonic tank is (45 × 30 × 30 in cm) 40 L. A liquid level corresponding to 28.35 L is maintained for all experiments. Hence the corresponding quantity of coal to be processed is 2.835 kg (1:10). As per the previous investigation, reagent-based ultrasonic methods remove more than 90% of maximum total sulfur even at the lowest coal-to-reagent (1: 3) volumes included in this investigation. Hence, 9.45 kg of coal can be processed using 28.35 L of 3 volume % of H_2O_2 during 30 min of treatment time. From Table 5.8, it may be observed that the coal processed per day using lab-scale ultrasonic tank is equivalent to approximately 0.5 ton/day. The reagent consumed for processing the 0.5 ton of coal is 1,360 L of 3 volume % H_2O_2.

Table 5.8 Amount of coal that can be processed per day using lab scale ultrasonic tank

Coal-reagent ratio (g/cc)	Mass of coal (g)	Volume of reagent (ml)	Size of container (L)	Treatment time	TSR, %
1:10	10	100	0.25	30 min	93.4
1:10	50	500	1		85.21
1:10	150	1,500	3		87.94
1:10	2,835	28,350	40		–
1:03	9,450	28,350	[(45×30×30)		
	18,900	56,700	in cm]	1 h	
	453,600	1,360,800		1 day	
Kg per day	**453.60**	**1,360.8 L**	**40**	**1 day**	

5.7 Summary

- In 25 kHz system, the total sulfur removal occurs mainly due to cavitation mechanism, and streaming mechanism causes removal in 430 kHz system. Dual frequency renders highest removal due to combined mechanisms. This has been confirmed by observing fluid behavior in ultrasonic tank, and by the size distribution analysis of virgin and sonicated coal.
- In almost all cases, higher reagent concentration yields higher removal. Total sulfur removal from coal is a linear function of sonication time, increasing with increasing sonication time.
- Particle breakage increases with decreasing particle size. It may be observed that removal occurs due to particle breakage as well as diffusional effect in the case of $-600 + 212$ microns size-range of coal particle.
- The optimum set of conditions determined from the investigation was validated with different types of coal, showing good removal efficiencies for sulfur as well as for ash.
- A mechanistic model for ultrasonic reagent-based coal desulfurization was developed. This model yields significant parameters for each reagent.
- Coal processed per day using lab-scale ultrasonic tank is equivalent to approximately 0.5 ton/day. The reagent consumed for processing the 0.5 ton of coal is 1,360 L of 3 volume % H_2O_2.

The present investigation shows that, reagent-based ultrasonic coal wash appears to be a promising technique to remove sulfur and ash from coal. Results indicated that more than 90% removal was achieved within the bounds of low concentration, low treatment time and low reagent volume consumption. This has a positive implication for scaling up reagent-based ultrasonic coal wash to larger coal quantities. The different types of coal tested using optimum conditions derived from the investigation have yielded good results, increasing the level of confidence in scale-up efforts for ultrasonic reagent-based coal wash. This method has the potential to replace conventional methods in terms of less treatment time, less reagent volume, low reagent concentration and commercially-available ultrasonic coal-wash equipment. At this stage, laboratory-scale results are promising enough that a larger-scale trial with high-sulfur coals is strongly recommended.

References

1. Ratoarinoro N, Contamlne F, Wilhelm AM, Berlan J, Delmas H (1995) Activation of a solid–liquid chemical reaction by ultrasound. Chem Eng Sci 50:554–558
2. Sano Y, Yamagushi N, Adachi T (1974) Mass transfer coefficients for suspended particles in agitated vessels and bubble columns. J Chem Eng Jpn 7:255
3. Mazet N (1992) Modeling of gas-solid reactions. 1. Nonporous solids. Int J Chem Engg 32:271–284

Chapter 6
Assessment of Benefits from Ultrasonic Coal-Wash (USCW)

6.1 Introduction

Two major problems arising from the burning of high-ash and high-sulfur coal are; (1) erosion of boiler accessories, and (2) environment and health hazards. The mitigation of these is the main objective of the present study, by employing ultrasonic coal-wash on high-ash and high-sulfur coal. This method can be used to control particulate and sulfur emissions prior to coal combustion. This investigation (USCW) has been primarily focused on these issues. Outcome of these studies has been presented earlier in this thesis. The issue of scale-up to production quantities will be discussed in this chapter.

6.2 Proposed Flow Chart for USCW on Industrial Scale

A method currently employed in production is based on "Vibrating Tray". A schematic view of the Vibrating Tray setup is shown in Fig. 6.1 (www.advancedsonics.com).

The Vibrating Tray equipment is a high volume ultrasonic trough which is effective in accelerating the surface dynamics of the fluidized particles. The Tray provides a large volumetric capacity for materials that benefit from mild acoustic exposure. The ultrasonic cavitational energy scrubs the surface of each particle as well as its interior part as it flows over the Tray. The cleaning effect produced by water alone is very effective in removing surface contaminates from the particulate pores. Chemical additives, added prior to the ultrasonic Vibrating Tray, become highly reactive in the acoustic field.

On the basis of laboratory-scale experimental investigation and results, a flow chart for USCW on an industrial scale is proposed here. This can be implemented as a continuous or batch process. Even in such a large scale, volume and concentration of H_2O_2 can be minimized while still achieving high efficiency of ash

B. Ambedkar, *Ultrasonic Coal-Wash for De-Ashing and De-Sulfurization*, Springer Theses, DOI: 10.1007/978-3-642-25017-0_6,

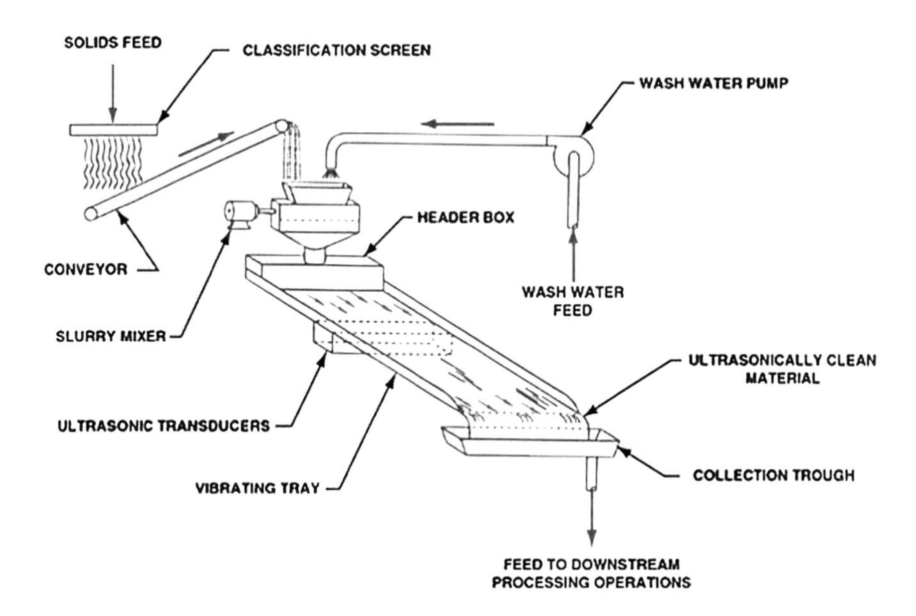

Fig. 6.1 Schematic view of vibrating trays

and sulfur removal (as evident from Figs. 5.2, 5.5). This process can potentially produce thousands of tons of washed coal per day. Figure 6.2 shows the proposed flow diagram for USCW.

Run-of-mine coal is crushed to obtain smaller sized coal particles from very big lumps. Then, the crushed coal particles are used to prepare coal slurry. Coal slurry preparation can be aqueous or reagent-based. The prepared coal slurry is subjected to ultrasonication. Based on physical property of the coal, low, high or combined frequency (low and high) may be chosen for ultrasonication. The sonicated coal slurry may be passed through a settling column, and sufficient time provided for the particles to settle. Coal, being lighter than ash material, will have low settling velocity and will settle slower, facilitating separation. Various grades (ash content) of coal sample can be collected with respect to the height of the decanting column. Otherwise, series of hydro cyclones can be put into operation for separating ash-free coal from mixture of coal and ash-rich coal. The heavier ash-rich coal is recycled for further sonication to get ash-free coal. Then, the separated coal sample is subjected to surfactant wash to remove further impurities which are adhered or gently sticking on the coal surface. Surfactant wash, followed by two consecutive water-washes is required to neutralize the reagent or surfactant-doped coal sample. Ultrasonic method is more effective in neutralizing the reagent-doped coal sample compared to conventional coal-wash. Surfactant and water-wash may therefore be carried out in presence of non-erosive high-frequency ultrasound. Finally, the neutralized coal sample may be dried and taken for further processing.

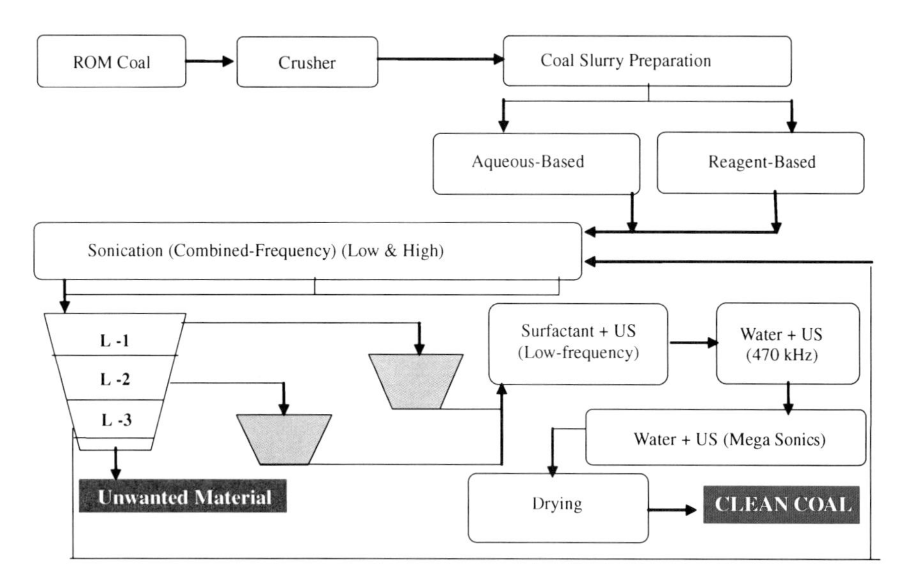

Fig. 6.2 Proposed flow diagram for ultrasonic method of coal-wash

6.3 Effect of USCW on Metal Erosion and Corrosion Due to Burning of High-Ash and Sulfur Coal

Erosion of coal burning boilers is caused by a combination of physical and chemical effects. From the fly-ash impact erosion study, particle size and quantity of fly ash (Eq. 3.3) were identified as key factors in determining the rate of metal erosion. This led to the investigation of low-frequency, low-intensity ultrasonics for size reduction and washing of high-ash coals. Figure 6.3 shows the projected effect of USCW on fly-ash impact metal erosion, accounting for the associated reduction in particle size and ash content. Under optimum ultrasonic coal-wash conditions, the associated size reduction of washed coal is 30% and the corresponding average ash removal is about 22% (Table 4.8). Hence the reduced erosion due to coal particle size reduction and ash removal is potentially more than 50%.

It has been reported that this biomass-fired boiler fly-ash has relatively high erosion due to its composition containing high concentrations of chemically active compounds of alkali, sulfur, phosphorus and chlorine [1]. Among these, sulfur and chlorine are the key components to stimulate corrosion-accelerated erosion. The coal tested here was not from the coastal region; and hence, chlorine is excluded from our study. Burning of high-sulfur coal leaves part of sulfur in form of SO_x gas which pollutes the environment, and part of it in chemical combination with the other elements present in the fly-ash to form sulphates of calcium, sodium, potassium, magnesium, aluminum, etc. These are highly corrosive, adhering strongly on the surface of the boiler accessories and accelerating the corrosion. This damage is more severe at high temperature. The major mechanism of material

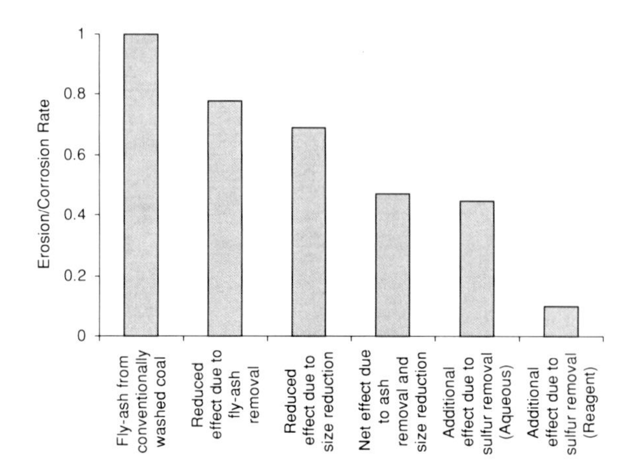

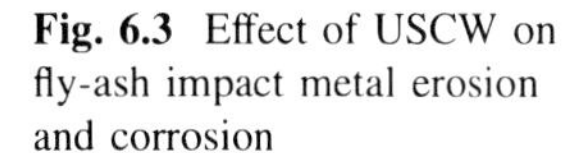
Fig. 6.3 Effect of USCW on fly-ash impact metal erosion and corrosion

wastage is therefore erosion-stimulated corrosion and corrosion-accelerated erosion. From the de-sulfurization studies, more than 90% TSR using reagents-based ultrasonics and 50–60% of TSR was observed using aqueous-based ultrasonic coal de-sulfurization, and the associated reduction in corrosion-accelerated erosion is shown in Fig. 6.3. Simultaneous de-ashing and de-sulfurization of coals using ultrasonics has the potential to significantly mitigate this problem.

6.4 Particulate and SO_X Emission

Mined coal is of variable quality and is frequently associated with mineral and chemical material including clay, sand, sulfur and trace elements. Particulate emissions are finely-divided solid and liquid (other than water) substances that are emitted from coal-fired power stations. A number of technologies have been developed to control particulate emissions and are widely deployed in both developed and developing countries, including: electrostatic precipitators, fabric filters or bag houses, wet particulate scrubbers and hot gas filtration systems. These are the methods which can control particulate emissions during or post-combustion. Coal cleaning by washing and beneficiation removes this associated material, prepares the coal to customer specifications and is an important step in reducing emissions from coal use. Ultrasonic method of coal-wash can reduce the particulate emissions prior to coal combustion by removing ash material from it, thereby reducing transportation, storage and handling costs. Effect of high-sulfur coal burning and methods that are in current practice to control SO_x emission, as well as the drawbacks associated with these methods are discussed in Sect. 1.1. Aqueous-based ultrasonic coal de-sulfurization results in nearly complete removal of sulphate sulfur, and part of pyritic and organic sulfur, as discussed in Sect. 4.4.3. In the case of reagent-based coal de-sulfurization, greater than 90% total sulfur removal is possible, as discussed in Chap. 5. This will reduce most of the environmental and health hazards. Hence, ultrasonic coal cleaning reduces the ash content of coal resulting in less waste, lower sulfur dioxide (SO_2) emissions and improved thermal efficiencies, leading to lower CO_2 emissions.

Reference

1. Wang BQ (1995) Erosion-corrosion of coatings by bio-mass-fired boiler fly ash. Wear 188:40–48

Chapter 7
Summary, Conclusions and Recommendations

The present study has brought out several interesting observations and fundamental insights on fly-ash impact erosion and ultrasonic coal-wash which are very useful while scaling up the process to larger coal quantities. Some of the key conclusions of the present work are:

- Effectiveness of Ultrasonic coal-wash for Indian coals has been demonstrated.
- Ultrasound-assisted coal particle breakage mechanism has been proposed.
- Effect of several parameters has been studied systematically.
- Dual-mode frequency (low and high) of application has been found to be the most effective.
- For the removal of sulfate sulfur, aqueous-based US is found to be adequate.
- For total sulfur removal, H_2O_2-based coal washing is found to be the best from the process efficiency as well as environmental compatibility point of view.
- Optimal parameters for the process have been determined experimentally, and scale-up studies up to a factor of 15 have been carried out.
- Based on these studies, a continuous coal wash programme for simultaneous ash and sulfur removal has been developed.

7.1 Recommendations for Future Work

- Effect of shape of the impacting particle on metal erosion needs to be studied.
- Mass conservation on decanted coal samples at each level, and the corresponding size distribution, elemental and ash composition analyses need to be performed. This will provide a quantitative estimate of clean coal that can be obtained from ultrasonic de-ashing followed by decanting process.

B. Ambedkar, *Ultrasonic Coal-Wash for De-Ashing and De-Sulfurization*,
Springer Theses, DOI: 10.1007/978-3-642-25017-0_7,

- Further work is needed to interpret the similarities and differences between aqueous and reagent-based ultrasonic coal de-sulfurization.
- Economic analysis of the proposed process required.
- Coal-water slurry fuel (CWSF or CWS or CWF) is a fuel which consists of fine coal particles suspended in water. Presence of water in CWS reduces harmful emissions into the atmosphere, makes the coal explosion-proof, makes use of coal equivalent to use of liquid fuel (e.g., heating oil), and gives other benefits. Such slurry can be prepared using power ultrasound. The slurry behavior needs to be studied to maintain the suspension for a longer period.
- Surface of the sonicated coal has numerous pores and cracks leading to an increased burning rate as well as higher calorific value due to ash removal. Hence, effect of surface morphology on combustion behavior needs to be studied.

Based on laboratory experimental investigations, ultrasonic coal wash appears to be a promising technique to remove sulfur and ash from coal. Results indicated that considerable amount of ash and sulfur removal was achieved within the bounds of lower concentration, minimum treatment time and lowest reagent volume consumption. This has a positive implication for scaling up of ultrasonic coal-wash to larger coal quantities. The different types of coal tested using optimum conditions derived from the investigation have yielded good results, increasing the level of confidence in scale-up efforts for ultrasonic reagent-based coal wash. This method has the potential to replace conventional methods in terms of less treatment time, less reagent volume, low reagent concentration and commercially-available ultrasonic coal-wash equipment. At this stage, laboratory-scale results are promising enough that a larger-scale trial with high-ash and high-sulfur coal is strongly recommended.

Appendix I

Fly-ash impact metal erosion experimental datas

Tauguchi L_{27} orthoganol array Each trail was conducted twice							Grade 12		Grade 22		Grade A	
Trial No.	Velocity, (m/s)	Angle, degree	Size, microns	Ash type	Feed qty, (g)	Time, (min)	Aged coupon Avg. wt loss, (10^3 g)	New coupon Avg. wt loss, (10^3 g)	Aged coupon Avg. wt loss, (10^3 g)	New coupon Avg. wt loss, (10^3 g)	Aged coupon Avg. wt loss, (10^3 g)	New coupon Avg. wt loss, (10^3 g)
1	10	15	37.5	Mettur	5	5	0.15	0.35	0.35	0.20	0.35	0.15
2	10	15	112.5	Raichur	10	10	0.15	0.30	0.65	0.25	0.55	0.25
3	10	15	200	Dhahanu	15	15	0.2	0.25	0.40	0.30	0.35	0.10
4	10	30	37.5	Mettur	10	10	0.45	0.25	1.0	0.10	0.50	0.10
5	10	30	112.5	Raichur	15	15	0.55	0.50	0.90	0.20	0.65	0.30
6	10	30	200	Dhahanu	5	5	0.10	0.20	0.35	0.20	0.35	0.05
7	10	45	37.5	Mettur	15	15	0.35	0.20	0.60	0.30	0.35	0.10
8	10	45	112.5	Raichur	5	5	0.15	0.25	0.30	0.25	0.30	0.10
9	10	45	200	Dhahanu	10	10	0.10	0.20	0.35	0.20	0.25	0.10
10	20	15	37.5	Raichur	5	5	0.10	0.15	0.25	0.15	0.30	0.075
11	20	15	112.5	Dhahanu	10	10	0.40	0.35	0.55	0.25	0.40	0.30
12	20	15	200	Mettur	15	15	0.95	0.55	1.10	0.65	1.5	0.15
13	20	30	37.5	Raichur	10	10	0.10	0.20	0.30	0.25	0.30	0.05
14	20	30	112.5	Dhahanu	15	15	0.50	0.30	0.45	0.40	0.60	0.25
15	20	30	200	Mettur	5	5	0.30	0.25	0.70	0.30	0.60	0.10
16	20	45	37.5	Raichur	15	15	0.15	0.30	0.30	0.30	0.50	0.10
17	20	45	112.5	Dhahanu	5	5	0.25	0.15	0.30	0.30	0.30	0.15
18	20	45	200	Mettur	10	10	0.90	0.65	1.30	0.15	1.20	0.075

(continued)

B. Ambedkar, *Ultrasonic Coal-Wash for De-Ashing and De-Sulfurization*,
Springer Theses, DOI: 10.1007/978-3-642-25017-0,

(continued)

Tauguchi L_{27} orthoganol array Each trail was conducted twice							Grade 12		Grade 22		Grade A	
							Aged coupon	New coupon	Aged coupon	New coupon	Aged coupon	New coupon
Trial No.	Velocity, (m/s)	Angle, degree	Size, microns	Ash type	Feed qty, (g)	Time, (min)	Avg. wt loss, (10^3 g)	Avg. wt loss, (10^3 g)	Avg. wt loss, (10^3 g)	Avg. wt loss, (10^3 g)	Avg. wt loss, (10^3 g)	Avg. wt loss, (10^3 g)
19	30	15	37.5	Dhahanu	5	5	0.20	0.15	0.30	0.15	0.25	0.05
20	30	15	112.5	Mettur	10	10	0.90	0.90	1.50	0.85	2.15	0.30
21	30	15	200	Raichur	15	15	0.55	0.60	1.25	0.40	1.60	0.15
22	30	30	37.5	Dhahanu	10	10	0.20	0.25	0.35	0.20	0.40	0.05
23	30	30	112.5	Mettur	15	15	1.0	0.70	1.45	1.0	2.20	0.55
24	30	30	200	Raichur	5	5	0.20	0.20	0.50	0.45	0.75	0.075
25	30	45	37.5	Dhahanu	15	15	0.20	0.20	0.45	0.30	0.40	0.075
26	30	45	112.5	Mettur	5	5	0.20	0.35	0.75	0.50	1.05	0.20
27	30	45	200	Raichur	10	10	0.35	0.40	0.70	0.30	1.10	0.15

Appendix II

Data analysis	
Effect of impact velocity on erosion	10 m/s = Average of (1 + 2 + 3 + 4 + 5 + 6 + 7 + 8+9) trial erosion data points
	20 m/s = Average of (10 + 11 + 12 + 13 + 14 + 15 + 16 + 17 + 18) trial erosion data points
	30 m/s = Average of (19 + 20 + 21 + 22 + 23 + 24 + 25 + 26 + 27) trial erosion data points
Effect of impact angle on erosion	15° = Average of (1 + 2 + 3 + 10 + 11 + 12 + 19 + 20 + 21) trial erosion data points
	30° = Average of (4 + 5 + 6 + 13 + 14 + 15 + 22 + 23 + 24) trial erosion data points
	45° = Average of (7 + 8 + 9 + 16 + 17 + 18 + 25 + 26 + 27) trial erosion data points
Effect of fly-ash particle size on erosion	37.5 μm = Average of (1 + 4 + 7 + 10 + 13 + 16 + 19 + 22 + 25) trial erosion data points
	112.5 μm = Average of (2 + 5 + 8 + 11 + 14 + 17 + 20 + 23 + 26) trial erosion data points
	200 μm = Average of (3 + 6 + 9 + 12 + 15 + 18 + 21 + 24 + 27) trial erosion data points
Effect of ash type on erosion	Mettur = Average of (1 + 4 + 7 + 12 + 15 + 18 + 20 + 23 + 26) trial erosion data points
	Raichur = Average of (2 + 5 + 8 + 10 + 13 + 16 + 21 + 24 + 27) trial erosion data points
	Dhahanu = Average of (3 + 6 + 9 + 11 + 14 + 17 + 19 + 22 + 25) trial erosion data points
Effect of fly-ash quantity on erosion	5 g = Average of (1 + 6 + 8 + 10 + 15 + 17 + 19 + 24 + 26) trial erosion data points
	10 g = Average of (2 + 4 + 9 + 11 + 13 + 18 + 20 + 22 + 27) trial erosion data points
	15 g = Average of (3 + 5 + 7 + 12 + 14 + 16 + 21 + 23 + 25) trial erosion data points
Effect of time on erosion	5 min = Average of (1 + 6 + 8 + 10 + 15 + 17 + 19 + 24 + 26) trial erosion data points
	10 min = Average of (2 + 4 + 9 + 11 + 13 + 18 + 20 + 22 + 27) trial erosion data points
	15 min = Average of (3 + 5 + 7 + 12 + 14 + 16 + 21 + 23 + 25) trial erosion data points
The above analysis has also been used in the following Appendix	

B. Ambedkar, *Ultrasonic Coal-Wash for De-Ashing and De-Sulfurization*,
Springer Theses, DOI: 10.1007/978-3-642-25017-0,

Appendix III

Reagent (Methanol)—based de-ashing and size reduction experimental data's for Belphar coal

							Ash analysis			
Frequency, (KHz)	Initial coal size, (mm)	Coal-solvent ratio, vol (%)	Initial temp, (C)	Coal type	Initial wt. of coal, (g)	Final wt. of coal, (g)	Initial dish weight, (g)	Dish + sam weight, (g)	Dish + ash weight, (g)	Dish wt.after brushing, (g)
25	(− 1 + 0.09)	1 : 6	25	Reject	10	8.06	16.47	16.95	16.69	16.47
					10	8.19	17.53	18.03	17.75	17.53
25	(− 1 + 0.09)	1 : 6	25	Washed	10	8.04	16.79	17.27	16.95	16.79
					10	8.23	16.94	17.43	17.11	16.94
25	(− 1 + 0.09)	1 : 6	25	ROM	10	8.16	17.2	17.72	17.4	17.2
					10	8.42	18.03	18.5	18.2	18.03
25	(− 2 + 1)	1 : 4	35	Reject	10	9.11	16.46	16.95	16.68	16.46
					10	8.95	17.53	18.04	17.77	17.53
25	(− 2 + 1)	1 : 4	35	Washed	10	8.86	16.79	17.28	16.96	16.79
					10	8.76	16.93	17.42	17.1	16.96
25	(− 2 + 1)	1 : 4	35	ROM	10	9.15	17.2	17.7	17.41	17.2
					10	8.99	18.03	18.52	18.23	18.03
25	(− 4 + 2)	1 : 2	45	Reject	10	9.25	16.47	16.97	16.73	16.47
					10	9.01	17.53	18.02	17.78	17.53

(continued)

B. Ambedkar, *Ultrasonic Coal-Wash for De-Ashing and De-Sulfurization*,
Springer Theses, DOI: 10.1007/978-3-642-25017 0,

(continued)

							Ash analysis			
Frequency, (KHz)	Initial coal size, (mm)	Coal-solvent ratio, vol (%)	Initial temp, (C)	Coal type	Initial wt. of coal, (g)	Final wt. of coal, (g)	Initial dish weight, (g)	Dish + sam weight, (g)	Dish + ash weight, (g)	Dish wt.after brushing, (g)
25	(− 4 + 2)	1 : 2	45	Washed	10	8.56	16.8	17.29	16.98	16.8
					10	8.67	16.94	17.44	17.14	16.94
25	(− 4 + 2)	1 : 2	45	ROM	10	8.78	17.2	17.7	17.44	16.2
					10	8.63	18.02	18.53	18.26	18.02
132	(− 1 + 0.09)	1 : 4	45	Washed	10	8.79	16.48	16.98	16.65	16.48
					10	8.63	17.53	18	17.68	17.53
132	(− 1 + 0.09)	1 : 4	45	ROM	10	8.38	16.78	17.28	16.98	16.78
					10	8.44	17.2	17.7	17.4	17.2
132	(− 1 + 0.09)	1 : 4	45	Reject	10	8.98	18.03	18.53	18.26	18.03
					10	8.79	16.95	17.44	17.16	16.95
132	(− 2 + 1)	1 : 2	25	Washed	10	9.03	16.46	17.00	16.68	16.46
					10	8.91	17.53	18.04	17.74	17.53
132	(− 2 + 1)	1 : 2	25	ROM	10	9.04	16.79	17.30	17.00	16.79
					10	9.21	17.20	17.71	17.40	17.20
132	(− 2 + 1)	1 : 2	25	Reject	10	9.27	16.95	17.45	17.19	16.95
					10	9.11	18.02	18.51	18.27	18.02
132	(− 4 + 2)	1 : 6	35	ROM	10	8.66	16.48	16.97	16.69	16.48
					10	8.5	17.52	18.04	17.76	17.52
132	(− 4 + 2)	1 : 6	35	Reject	10	8.87	16.80	17.30	17.06	16.80
					10	8.6	16.93	17.43	17.17	16.93
132	(− 4 + 2)	1 : 6	35	Washed	10	8.51	17.19	17.69	17.39	17.19
					10	8.42	18.03	18.52	18.22	18.03
470	(− 1 + 0.09)	1 : 2	35	ROM	10	8.62	16.48	16.98	16.65	16.48
					10	8.8	17.53	18.03	17.73	17.53

(continued)

(continued)

							Ash analysis			
Frequency, (KHz)	Initial coal size, (mm)	Coal-solvent ratio, vol (%)	Initial temp, (C)	Coal type	Initial wt. of coal, (g)	Final wt. of coal, (g)	Initial dish weight, (g)	Dish + sam weight, (g)	Dish + ash weight, (g)	Dish wt.after brushing, (g)
470	(− 1 + 0.09)	1 : 2	35	Reject	10	8.71	16.80	17.28	16.99	16.80
					10	8.97	16.94	17.45	17.17	16.94
470	(− 1 + 0.09)	1 : 2	35	Washed	10	8.59	17.20	17.68	17.33	17.20
					10	8.69	18.03	18.53	18.18	18.03
470	(− 2 + 1)	1 : 6	45	ROM	10	9.03	16.47	16.99	16.68	16.47
					10	9.17	17.54	18.03	17.73	17.54
470	(− 2 + 1)	1 : 6	45	Reject	10	8.92	16.78	17.29	17.04	16.78
					10	9.1	16.94	17.44	17.17	16.94
470	(− 2 + 1)	1 : 6	45	Washed	10	8.86	17.18	17.72	17.38	17.18
					10	8.8	18.04	18.52	18.21	18.04
470	(− 4 + 2)	1 : 4	25	ROM	10	9.11	16.47	16.97	16.68	16.47
					10	8.8	17.54	18.02	17.75	17.54
470	(− 4 + 2)	1 : 4	25	Reject	10	8.69	16.78	17.27	17.04	16.78
					10	8.88	16.98	17.44	17.16	16.98
470	(− 4 + 2)	1 : 4	25	Washed	10	9.19	17.23	17.72	17.39	17.23
					10	8.91	18.02	18.53	18.22	18.02

Appendix IV

Reagent (Methanol)—based de-ashing and size reduction experimental data's for Dipka coal

							Ash analysis			
Frequency, (KHz)	Initial coal size, (mm)	Coal-solvent ratio, vol (%)	Initial temp, (C)	Coal type	Initial wt. of coal, (g)	Final wt. of coal, (g)	Initial dish weight, (g)	Dish + sam weight, (g)	Dish + ash weight, (g)	Dish wt.after brushing, (g)
25	(− 1 + 0.09)	1 : 6	25	Reject	10	8.68	16.48	16.99	16.77	16.48
					10	8.79	17.53	18.03	17.8	17.53
25	(− 1 + 0.09)	1 : 6	25	Washed	10	8.47	16.79	17.28	16.89	16.79
					10	8.33	16.94	17.35	17.02	16.94
25	(− 1 + 0.09)	1 : 6	25	ROM	10	8.81	17.2	17.71	17.3	17.2
					10	8.7	18.03	18.52	18.12	18.03
25	(− 2 + 1)	1 : 4	35	Reject	10	9.24	16.47	16.98	16.76	16.47
					10	9.36	17.53	18.04	17.85	17.53
25	(− 2 + 1)	1 : 4	35	Washed	10	8.7	16.79	17.29	16.94	16.79
					10	8.81	16.92	17.44	17.06	16.92
25	(− 2 + 1)	1 : 4	35	ROM	10	8.69	17.2	17.7	17.29	17.2
					10	8.77	18.03	18.55	18.13	18.03
25	(− 4 + 2)	1 : 2	45	Reject	10	9.38	16.94	17.45	17.24	16.94
					10	9.23	17.54	18.05	17.85	17.54
25	(− 4 + 2)	1 : 2	45	Washed	10	8.57	16.79	17.31	16.94	16.79
					10	8.7	16.47	16.98	16.63	16.47
25	(− 4 + 2)	1 : 2	45	ROM	10	8.44	18.03	18.55	18.23	18.03
					10	8.57	17.2	17.71	17.39	17.2

(continued)

B. Ambedkar, *Ultrasonic Coal-Wash for De-Ashing and De-Sulfurization*,
Springer Theses, DOI: 10.1007/978-3-642-25017-0,

(continued)

							Ash analysis			
Frequency, (KHz)	Initial coal size, (mm)	Coal-solvent ratio, vol (%)	Initial temp, (C)	Coal type	Initial wt. of coal, (g)	Final wt. of coal, (g)	Initial dish weight, (g)	Dish + sam weight, (g)	Dish + ash weight, (g)	Dish wt.after brushing, (g)
132	(− 1 + 0.09)	1 : 4	45	Washed	10	8.76	16.94	17.45	17.09	16.94
					10	8.83	17.53	18.03	17.69	17.53
132	(− 1 + 0.09)	1 : 4	45	ROM	10	8.49	16.76	17.31	16.94	16.76
					10	8.4	16.48	16.98	16.61	16.48
132	(− 1 + 0.09)	1 : 4	45	Reject	10	9.3	18.04	18.55	18.36	18.04
					10	9.17	17.2	17.72	17.52	17.2
132	(− 2 + 1)	1 : 2	25	Washed	10	8.7	16.48	17	16.62	16.48
					10	8.83	18.03	18.53	18.19	18.03
132	(− 2 + 1)	1 : 2	25	ROM	10	8.72	17.2	17.71	17.32	17.2
					10	8.81	16.93	17.44	17.07	16.93
132	(− 2 + 1)	1 : 2	25	Reject	10	9.6	17.53	18.04	17.85	17.53
					10	9.39	16.78	17.29	17.12	16.78
132	(− 4 + 2)	1 : 6	35	ROM	10	8.6	16.48	16.97	16.62	16.48
					10	8.53	18.03	18.55	18.2	18.03
132	(− 4 + 2)	1 : 6	35	Reject	10	9.1	17.19	17.7	17.5	17.19
					10	8.9	16.94	17.44	17.25	16.94
132	(− 4 + 2)	1 : 6	35	Washed	10	8.25	17.53	18.05	17.72	17.53
					10	8.43	16.79	17.31	16.97	16.79
470	(− 1 + 0.09)	1 : 2	35	ROM	10	9.07	16.47	16.96	16.58	16.47
					10	9.22	18.03	18.56	18.15	18.03
470	(−1 + 0.09)	1 : 2	35	Reject	10	9.6	17.18	17.71	17.53	17.18
					10	9.34	16.94	17.46	17.25	16.94
470	(− 1 + 0.09)	1 : 2	35	Washed	10	8.25	17.54	18.04	17.65	17.54
					10	8.44	16.79	17.28	16.91	16.79

(continued)

(continued)

					Ash analysis					
Frequency, (KHz)	Initial coal size, (mm)	Coal-solvent ratio, vol (%)	Initial temp, (C)	Coal type	Initial wt. of coal, (g)	Final wt. of coal, (g)	Initial dish weight, (g)	Dish + sam weight, (g)	Dish + ash weight, (g)	Dish wt.after brushing, (g)
470	(− 2 + 1)	1 : 6	45	ROM	10	8.81	16.47	16.97	16.59	16.47
					10	8.67	18.03	18.57	18.16	18.03
470	(− 2 + 1)	1 : 6	45	Reject	10	9.58	17.2	17.71	17.51	17.2
					10	9.33	16.94	17.45	17.24	16.94
470	(− 2 + 1)	1 : 6	45	Washed	10	8.76	17.54	18.03	17.65	17.54
					10	8.9	16.78	17.28	16.91	16.78
470	(− 4 + 2)	1 : 4	25	ROM	10	8.5	16.46	16.96	16.62	16.46
					10	8.25	18.03	18.53	18.18	18.03
470	(− 4 + 2)	1 : 4	25	Reject	10	9.27	17.21	17.71	17.52	17.21
					10	9.04	16.94	17.45	17.28	16.94
470	(− 4 + 2)	1 : 4	25	Washed	10	8.68	17.53	18.05	17.67	17.53
					10	9	16.79	17.3	16.92	16.79

Appendix V

Reagent (Methanol)—based de-ashing and size reduction experimental data's for Talcher coal

							Ash analysis			
Frequency, (KHz)	Initial coal size, (mm)	Coal-solvent ratio, vol (%)	Initial temp, (C)	Coal type	Initial wt. of coal, (g)	Final wt. of coal, (g)	Initial dish weight, (g)	Dish + sam weight, (g)	Dish + ash Weight, (g)	Dish wt.after brushing, (g)
25	(− 1 + 0.09)	1 : 6	25	Reject	10.05	8.51	17.19	17.71	17.45	17.19
					10.05	8.48	17.53	18.05	17.79	17.53
25	(− 1 + 0.09)	1 : 6	25	Washed	10.05	8.59	16.79	17.29	16.89	16.79
					10.05	8.61	18.02	18.54	18.09	18.03
25	(− 1 + 0.09)	1 : 6	25	ROM	10.05	8.5	17.2	17.71	17.31	17.2
					10.05	8.41	17.53	18.02	17.64	17.53
25	(− 2 + 1)	1 : 4	35	Reject	10.05	8.93	16.79	17.3	17.05	16.79
					10.05	8.79	18.02	18.53	18.29	18.02
25	(− 2 + 1)	1 : 4	35	Washed	10.02	8.71	17.2	17.71	17.3	17.19
					10.04	8.33	17.54	18.05	17.64	17.53
25	(− 2 + 1)	1 : 4	35	ROM	10.05	8.82	16.79	17.3	16.94	16.79
					10.07	9.01	18.03	18.54	18.18	18.02
25	(− 4 + 2)	1 : 2	45	Reject	10.03	8.98	17.2	17.68	17.48	17.19
					10.04	9.1	18.03	18.52	18.29	18.02
25	(− 4 + 2)	1 : 2	45	Washed	10.04	8.8	17.52	18.05	17.65	17.52
					10.05	8.72	16.78	17.3	16.92	16.78
25	(− 4 + 2)	1 : 2	45	ROM	10.02	8.69	17.2	17.68	17.39	17.2
					10.02	8.73	18.02	18.48	18.23	18.02

(continued)

B. Ambedkar, *Ultrasonic Coal-Wash for De-Ashing and De-Sulfurization*,
Springer Theses, DOI: 10.1007/978-3-642-25017-0,

(continued)

							Ash analysis			
Frequency, (KHz)	Initial coal size, (mm)	Coal-solvent ratio, vol (%)	Initial temp, (C)	Coal type	Initial wt. of coal, (g)	Final wt. of coal, (g)	Initial dish weight, (g)	Dish + sam weight, (g)	Dish + ash Weight, (g)	Dish wt.after brushing, (g)
132	(− 1 + 0.09)	1 : 4	45	Washed	10.04	8.71	17.54	18.05	17.61	17.53
					10.06	8.83	16.79	17.29	16.87	16.79
132	(− 1 + 0.09)	1 : 4	45	ROM	10.03	8.74	17.2	17.7	17.31	17.2
					10.01	8.7	18.03	18.55	18.14	18.03
132	(− 1 + 0.09)	1 : 4	45	Reject	10.02	8.98	17.53	18.01	17.76	17.53
					10.05	8.92	16.79	18.3	17.03	16.79
132	(− 2 + 1)	1 : 2	25	Washed	10.04	8.76	16.94	17.42	17.02	16.94
					10.04	8.79	16.47	16.95	16.56	16.47
132	(− 2 + 1)	1 : 2	25	ROM	10.02	8.56	16.46	16.95	16.58	16.46
					10.07	8.79	17.53	18.03	17.65	17.53
132	(− 2 + 1)	1 : 2	25	Reject	10.04	9.25	18.03	18.54	18.27	18.03
					10.01	9.19	16.94	17.43	17.17	16.94
132	(− 4 + 2)	1 : 6	35	ROM	10.12	8.74	16.78	17.26	16.88	16.78
					10.07	8.8	17.21	17.69	17.3	17.21
132	(− 4 + 2)	1 : 6	35	Reject	10.09	9.3	16.46	16.95	16.74	16.46
					10.06	9.15	17.52	18	17.8	17.52
132	(− 4 + 2)	1 : 6	35	Washed	10.06	8.56	18.03	18.53	18.18	18.03
					10.05	8.51	16.94	17.43	17.08	16.94
470	(− 1 + 0.09)	1 : 2	35	ROM	10.06	9.44	16.79	17.31	16.9	16.79
					10.06	9.31	17.2	17.69	17.3	17.2
470	(− 1 + 0.09)	1 : 2	35	Reject	10.07	9.4	16.46	16.93	16.69	16.46
					10.03	9.19	17.53	18.02	17.78	17.53
470	(− 1 + 0.09)	1 : 2	35	Washed	10.02	9.12	18.03	18.53	18.13	18.03
					10.02	9.15	16.92	17.39	17.02	16.92

(continued)

(continued)

					Ash analysis					
Frequency, (KHz)	Initial coal size, (mm)	Coal-solvent ratio, vol (%)	Initial temp, (C)	Coal type	Initial wt. of coal, (g)	Final wt. of coal, (g)	Initial dish weight, (g)	Dish + sam weight, (g)	Dish + ash Weight, (g)	Dish wt.after brushing, (g)
470	(− 2 + 1)	1 : 6	45	ROM	10.02	8.91	16.79	17.28	16.92	16.79
					10.02	9.11	17.2	17.7	17.34	17.2
470	(− 2 + 1)	1 : 6	45	Reject	10.07	9.59	16.47	16.95	16.72	16.47
					10.07	9.36	17.53	18.06	17.82	17.53
470	(− 2 + 1)	1 : 6	45	Washed	10.02	9.21	18.03	18.53	18.14	18.03
					10.03	8.93	16.94	17.44	17.04	16.94
470	(− 4 + 2)	1 : 4	25	ROM	10.08	8.91	16.78	17.28	16.9	16.78
					10.06	8.96	17.21	17.75	17.33	17.21
470	(− 4 + 2)	1 : 4	25	Reject	10.11	9.11	16.47	16.95	16.72	16.47
					10.09	8.98	17.54	18.03	17.8	17.54
470	(− 4 + 2)	1 : 4	25	Washed	10.06	9.21	16.94	17.44	17.04	16.94
					10.06	9.11	16.79	17.3	16.9	16.79

Appendix VI

Reagent (HNO_3)—based desulfurization experimental data's

Trail no.	Frequency, (kHz)	HNO_3 conc., (N)	Treatment time, (min)	Initial size, (μm)	Reagent volume, (ml)	Weight of residue, (g)	Total sulfur remaining, (%)
1	25	1	10	− 212 + 0	30	0.3376	3.9513
						0.3336	3.8963
2	25	1	20	− 600 + 212	60	0.2352	2.5443
						0.2303	2.4770
3	25	1	30	− 1,000 + 600	100	0.2052	2.1321
						0.1923	1.9549
4	25	3	10	− 212 + 0	30	0.3282	3.8221
						0.3187	3.6916
5	25	3	20	− 600 + 212	60	0.1908	1.9342
						0.2042	2.1184
6	25	3	30	− 1,000 + 600	100	0.2031	2.1033
						0.2039	2.1142
7	25	5	10	− 212 + 0	30	0.2752	3.0939
						0.2774	3.1241
8	25	5	20	− 600 + 212	60	0.1882	1.8985
						0.1681	1.6224
9	25	5	30	− 1000 + 600	100	0.1922	1.9535
						0.1721	1.6773
10	Dual	1	10	− 600 + 212	100	0.2121	2.2269
						0.1944	1.9837

(continued)

B. Ambedkar, *Ultrasonic Coal-Wash for De-Ashing and De-Sulfurization*,
Springer Theses, DOI: 10.1007/978-3-642-25017-0,

(continued)

Trail no.	Frequency, (kHz)	HNO_3 conc., (N)	Treatment time, (min)	Initial size, (μm)	Reagent volume, (ml)	Weight of residue, (g)	Total sulfur remaining, (%)
11	Dual	1	20	− 1,000 + 600	30	0.2102	2.2008
						0.1864	1.8738
12	Dual	1	30	− 212 + 0	60	0.2922	3.3275
						0.2692	3.0115
13	Dual	3	10	− 600 + 212	100	0.1712	1.6649
						0.1885	1.9026
14	Dual	3	20	− 1,000 + 600	30	0.2162	2.2832
						0.2285	2.4522
15	Dual	3	30	− 212 + 0	60	0.2592	2.8741
						0.2542	2.8054
16	Dual	5	10	− 600 + 212	100	0.2042	2.1184
						0.1819	1.8120
17	Dual	5	20	− 1,000 + 600	30	0.2117	2.2214
						0.1837	1.8367
18	Dual	5	30	− 212 + 0	60	0.2549	2.8150
						0.2312	2.4893
19	430	1	10	− 1,000 + 600	60	0.2367	2.5649
						0.2269	2.4303
20	430	1	20	− 212 + 0	100	0.2984	3.4127
						0.2804	3.1654
21	430	1	30	− 600 + 212	30	0.2133	2.2434
						0.1824	1.8188
22	430	3	10	− 1,000 + 600	60	0.2221	2.3643
						0.2288	2.4564
23	430	3	20	− 212 + 0	100	0.2757	3.1008
						0.2661	2.9689

(continued)

(continued)

Trail no.	Frequency, (kHz)	HNO_3 conc., (N)	Treatment time, (min)	Initial size, (μm)	Reagent volume, (ml)	Weight of residue, (g)	Total sulfur remaining, (%)
24	430	3	30	− 600 + 212	30	0.2057	2.1390
						0.2002	2.0634
25	430	5	10	− 1,000 + 600	60	0.2005	2.0675
						0.1922	1.9535
26	430	5	20	− 212 + 0	100	0.2394	2.6020
						0.2418	2.6350
27	430	5	30	− 600 + 212	30	0.1700	1.6485
						0.1964	2.0112

Appendix VII

Reagent (HNO_3)—based desulfurization experimental data's

Trail no.	Frequency, (kHz)	HNO_3 conc., (N)	Treatment time, (min)	Initial size, (μm)	Reagent volume, (ml)	Weight of residue, (g)	Total sulfur remaining, (%)
1	25	0.5	10	− 212 + 0	30	0.2232	2.368
						0.2313	2.480
2	25	0.5	20	− 600 + 212	60	0.1762	1.723
						0.1778	1.745
3	25	0.5	30	− 1,000 + 600	100	0.1998	2.047
						0.2171	2.285
4	25	1	10	− 212 + 0	30	0.1663	1.587
						0.1814	1.794
5	25	1	20	− 600 + 212	60	0.1405	1.232
						0.151	1.376
6	25	1	30	− 1,000 + 600	100	0.1956	1.989
						0.2098	2.184
7	25	2	10	− 212 + 0	30	0.1871	1.872
						0.2043	2.109
8	25	2	20	− 600 + 212	60	0.1086	0.794
						0.1236	1.000
9	25	2	30	− 1,000 + 600	100	0.0848	0.467
						0.0926	0.574
10	Dual	0.5	10	− 600 + 212	100	0.1685	1.617
						0.1556	1.440

(continued)

B. Ambedkar, *Ultrasonic Coal-Wash for De-Ashing and De-Sulfurization*,
Springer Theses, DOI: 10.1007/978-3-642-25017-0,

(continued)

Trail no.	Frequency, (kHz)	HNO_3 conc., (N)	Treatment time, (min)	Initial size, (μm)	Reagent volume, (ml)	Weight of residue, (g)	Total sulfur remaining, (%)
11	Dual	0.5	20	− 1,000 + 600	30	0.2185	2.304
						0.1915	1.933
12	Dual	0.5	30	− 212 + 0	60	0.2254	2.399
						0.2222	2.355
13	Dual	1	10	− 600 + 212	100	0.1666	1.591
						0.1725	1.672
14	Dual	1	20	− 1,000 + 600	30	0.1559	1.444
						0.1501	1.364
15	Dual	1	30	− 212 + 0	60	0.1603	1.504
						0.1537	1.414
16	Dual	2	10	− 600 + 212	100	0.1068	0.769
						0.1218	0.975
17	Dual	2	20	− 1,000 + 600	30	0.1753	1.710
						0.1504	1.368
18	Dual	2	30	− 212 + 0	60	0.0858	0.481
						0.1071	0.773
19	430	0.5	10	− 1,000 + 600	60	0.1877	1.881
						0.1946	1.975
20	430	0.5	20	− 212 + 0	100	0.152	1.390
						0.1466	1.316
21	430	0.5	30	− 600 + 212	30	0.148	1.335
						0.1458	1.305
22	430	1	10	− 1,000 + 600	60	0.1569	1.457
						0.1556	1.440
23	430	1	20	− 212 + 0	100	0.1977	2.018
						0.2145	2.249

(continued)

(continued)

Trail no.	Frequency, (kHz)	HNO_3 conc., (N)	Treatment time, (min)	Initial size, (μm)	Reagent volume, (ml)	Weight of residue, (g)	Total sulfur remaining, (%)
24	430	1	30	− 600 + 212	30	0.1675	1.603
						0.1815	1.795
25	430	2	10	− 1,000 + 600	60	0.1458	1.305
						0.1661	1.584
26	430	2	20	− 212 + 0	100	0.2126	2.223
						0.2324	2.495
27	430	2	30	− 600 + 212	30	0.1408	1.236
						0.1251	1.021

Appendix VIII

Reagent (H_2O_2)—based desulfurization experimental data's

Trail no.	Frequency, (kHz)	H_2O_2 conc., (N)	Treatment time, (min)	Initial size, (μm)	Reagent volume, (ml)	Weight of residue, (g)	Total sulfur remaining, (%)
1	25	0.5	10	− 212 + 0	30	0.1284	1.066
						0.1309	1.100
2	25	0.5	20	− 00 + 212	60	0.1108	0.824
						0.1086	0.794
3	25	0.5	30	− 1,000 + 600	100	0.1111	0.828
						0.1048	0.742
4	25	1	10	− 212 + 0	30	0.1496	1.357
						0.1473	1.326
5	25	1	20	− 600 + 212	60	0.0988	0.659
						0.0928	0.577
6	25	1	30	− 1,000 + 600	100	0.1185	0.930
						0.0917	0.562
7	25	2	10	− 212 + 0	30	0.144	1.280
						0.1077	0.781
8	25	2	20	-600 + 212	60	0.0511	0.004
						0.052	0.016
9	25	2	30	− 1,000 + 600	100	0.0595	0.119
						0.0722	0.294
10	Dual	0.5	10	− 600 + 212	100	0.1023	0.707
						0.1083	0.790

(continued)

B. Ambedkar, *Ultrasonic Coal-Wash for De-Ashing and De-Sulfurization*,
Springer Theses, DOI: 10.1007/978-3-642-25017-0,

(continued)

Trail no.	Frequency, (kHz)	H_2O_2 conc., (N)	Treatment time, (min)	Initial size, (μm)	Reagent volume, (ml)	Weight of residue, (g)	Total sulfur remaining, (%)
11	Dual	0.5	20	− 1,000 + 600	30	0.0872	0.500
						0.0921	0.567
12	Dual	0.5	30	− 212 + 0	60	0.1016	0.698
						0.092	0.566
13	Dual	1	10	− 600 + 212	100	0.0749	0.331
						0.0864	0.489
14	Dual	1	20	− 1,000 + 600	30	0.1254	1.025
						0.1212	0.967
15	Dual	1	30	− 212 + 0	60	0.0819	0.427
						0.1055	0.751
16	Dual	2	10	− 600 + 212	100	0.0555	0.064
						0.0531	0.031
17	Dual	2	20	− 1,000 + 600	30	0.1148	0.879
						0.1253	1.023
18	Dual	2	30	− 212 + 0	60	0.0943	0.597
						0.1005	0.683
19	430	0.5	10	− 1,000 + 600	60	0.1389	1.210
						0.135	1.157
20	430	0.5	20	− 212 + 0	100	0.1321	1.117
						0.1381	1.199
21	430	0.5	30	− 600 + 212	30	0.1132	0.857
						0.1151	0.883
22	430	1	10	− 1,000 + 600	60	0.1431	1.268
						0.1421	1.254
23	430	1	20	− 212 + 0	100	0.0965	0.628
						0.0885	0.518

(continued)

(continued)

Trail no.	Frequency, (kHz)	H_2O_2 conc., (N)	Treatment time, (min)	Initial size, (μm)	Reagent volume, (ml)	Weight of residue, (g)	Total sulfur remaining, (%)
24	430	1	30	− 600 + 212	30	0.1065	0.765
						0.1049	0.743
25	430	2	10	− 1,000 + 600	60	0.118	0.923
						0.12	0.950
26	430	2	20	− 212 + 0	100	0.095	0.607
						0.108	0.786
27	430	2	30	− 600 + 212	30	0.119	0.937
						0.139	1.212

Curriculum Vitae

1. **Name:** Ambedkar B

2. **Date of Birth:** 10th April 1979

3. **Educational Qualifications**

2001	**Bachelor of Technology (B.Tech)**
Institution:	Vellore Institute of Technology, Vellore
Specialization:	Chemical Engg.
2005	**Master of Technology (M.Tech)**
Institution:	National Institute of Technology, Trichy
Specialization:	Plant Design
2011	**Doctor of Philosophy (Ph.D)**
Institution:	Indian Institute of Technology, Chennai
Specialization:	Clean Coal Technology using Ultrasound

B. Ambedkar, *Ultrasonic Coal-Wash for De-Ashing and De-Sulfurization*,
Springer Theses, DOI: 10.1007/978-3-642-25017-0,